ServiceNow for Architects and Project Leaders

A complete guide to driving innovation, creating value, and making an impact with ServiceNow

Roy Justus

David Zhao

BIRMINGHAM—MUMBAI

ServiceNow for Architects and Project Leaders

Group Product Manager: Rahul Nair
Publishing Product Manager: Niranjan Naikwadi
Senior Editor: Arun Nadar
Content Development Editor: Sujata Tripathi
Technical Editor: Shruthi Shetty
Copy Editor: Safis Editing
Project Coordinator: Ashwin Dinesh Kharwa
Proofreader: Safis Editing
Indexer: Pratik Shirodkar
Production Designer: Roshan Kawale
Senior Marketing Coordinator: Nimisha Dua
Marketing Coordinator: Gaurav Christian

First published: December 2022
Production reference: 1181122

Published by Packt Publishing Ltd.
Livery Place
35 Livery Street
Birmingham
B3 2PB, UK.

ISBN 978-1-80324-529-4
www.packt.com

This book is possible thanks to Estella, Varisha, Allie, and our parents, whose support and understanding made this book possible, and thanks to all the team members, clients, and ServiceNow community members we've learned from along the way.

We dedicate this effort to those of you who are just starting your journey with ServiceNow so that you might walk an easier path than we did.

– Roy and David

Contributors

About the authors

Roy Justus has spent the last decade of his professional career helping to bring the potential of ServiceNow to life at scale. Within the ServiceNow ecosystem, he has been both a customer and a leader of consulting teams across a wide array of industries.

A lifelong learner, Roy holds a Bachelor of Commerce from the University of Calgary and is a **ServiceNow Certified Master Architect**. His current area of focus is on bringing the value of modern artificial intelligence to the ServiceNow ecosystem, both through the deployment of ServiceNow's AI features and the integration of external deep learning models.

David Zhao is a **ServiceNow Certified Master Architect** and has been obsessed with helping organizations bring business transformation to life through ServiceNow for nearly a decade. He has led complex transformations across a wide variety of industries, including energy and resources, telecom, financial services, and tech, across multiple continents and hemispheres.

Before getting involved in the ServiceNow world, David was once a web application developer and holds a master's degree in applied science from the University of Victoria with a focus on computer security research using statistical and AI models. He currently leads a team that focuses on advising organizations on the changes they must make to break through the *automation wall* that prevents them from taking full advantage of transformational technologies, such as ServiceNow's AI capabilities.

About the reviewer

Anubhav Kapoor has worked in the IT space for more than 13 years, with over 5 years dedicated explicitly to the ServiceNow platform. He has been an administrator, developer, and consultant and now works as an expert consultant on platform implementation and integration. He received his engineering degree from Dr. M.G.R. Educational and Research Institute in India. He is an active member and contributor to the ServiceNow community. When he is not at work, he is on a vacation with his family or blogging. He lives with his beautiful wife and two lovely daughters in New Jersey, USA.

My completion of this project could not have been accomplished without the support of my loving wife, Priyanka: my deepest gratitude. For taking care of the kids while their father completed his work, I could never be more thankful. Heartfelt thanks to my ServiceNow community members and experts for constantly putting in the best each day.

Table of Contents

3

Capturing Value from your Implementation   45

4

Planning an Implementation Program for Success   73

Part 2 – The Checklist

5

Securing Your ServiceNow Instances   101

Part 3 – From Success to Innovation

9

10

11

Designing Exceptional Experiences   205

Index   219

Other Books You May Enjoy   230

Preface

The work of a ServiceNow architect, project leader, or consultant is made meaningful by the impact it has on those who work with us and the users who ultimately benefit from our solutions. In creating this book, we hope to make the work of your teams more fulfilling and ultimately help you deliver on the promise of ServiceNow: to make the world of work, work better for people.

In this book, we share the approaches and perspectives that we've assembled over the last decade of our engagement in the ServiceNow ecosystem. We have tried to author the book we wish was available to us when we started our first project in the hope that it will make things easier, more efficient, and more effective for our readers and their teams.

This book will not replace the value of experience, but it will give you a head start and hopefully magnify your own ability to make a difference for those who are relying on your work. Thoughtful application of the tools and principles in this book will be rewarded as you deliver more impactful projects with fewer surprises.

Who this book is for

This book is targeted at architects, consultants, and project leaders looking to drive value by applying ServiceNow effectively and efficiently. Platform administration or development experience is useful but not required to get value from the book – however, some familiarity with the modules and features of ServiceNow is expected.

What this book covers

Chapter 1, Understanding the Value of ServiceNow, sets the stage by explaining why it is that we embark on ServiceNow projects in the first place. We will provide you with the necessary tools to frame the value proposition of your own project in consistent terms and make goals concrete enough to steer your actions.

Chapter 2, Recognizing and Avoiding Value Traps, outlines the common pitfalls that can put the value of your project at risk. We will explain why so many projects are affected by these traps and how to avoid them, as well as alternatives that help you stay out of trouble.

Chapter 3, Capturing Value from Your Implementation, will go over the multitude of ways that ServiceNow can add value to your organization, which will help you better identify and link opportunities together into strong business cases and set up the beneficiaries for success.

Chapter 4, Planning an Implementation Program for Success, provides you with a list of considerations and recommendations in case you are considering the implementation of ServiceNow in your organization, which can improve your value realization and the impact of your success.

Chapter 5, Securing Your ServiceNow Instances, is about making sure that your ServiceNow instance is protected from malicious actors. While anyone keeping up with industry news knows that cybersecurity breaches are increasingly impactful, there are steps you can take to make your instances less of a target.

Chapter 6, Managing Multiple ServiceNow Instances, intends to provide someone who is leading their first project with actionable guidelines on how to set up their ServiceNow instances, which will avoid them getting into trouble with releases.

Chapter 7, Designing Effective Processes at Scale, covers the capability of the ServiceNow platform to automate processes and how to connect the pieces to solve your automation problems and encourage configuration reuse.

Chapter 8, Platform Team Processes, Standards, and Techniques, provides an overview of some of the standards and norms a ServiceNow development team should establish in order to deliver value quickly and at a high quality.

Chapter 9, Effectively Operating ServiceNow, provides a lot of value if you are a ServiceNow platform owner. This chapter gives you a look at who you need on your team and which parts of the organization you should engage and provides other recommendations on making that team an effective business value enabler.

Chapter 10, Artificial Intelligence in ServiceNow, intends to provide coverage of one of the most important parts of the ServiceNow platform – the artificial intelligence features that can enable a reality of automation that many aspire to but few can deliver.

Chapter 11, Designing Exceptional Experiences, which is also the final chapter, is all about the ServiceNow user interface, explaining the recent changes to the platform, as well as making sense of the array of options that exist on it.

To get the most out of this book

This book is intended to provide value for practitioners at any stage of their ServiceNow journey but has been specifically written to be accessible for those starting their first ServiceNow project. While those with prior ServiceNow experience might be able to jump right into the chapters that seem most interesting, we recommend starting with *Part 1* as you frame your plan and business case for ServiceNow, picking up with *Part 2* before kicking off the technical configuration, and turning to *Part 3* before go-live or when you feel the topics might apply to you.

Download the color images

We also provide a PDF file that has color images of the screenshots and diagrams used in this book. You can download it here: `https://packt.link/qD8rj`.

Conventions used

There are a number of text conventions used throughout this book.

`Code in text`: Indicates code words in text, database table names, folder names, filenames, file extensions, pathnames, dummy URLs, user input, and Twitter handles. Here is an example: "To facilitate this distinction, it is often useful to name service accounts with a common prefix, such as `svc-`."

Bold: Indicates a new term, an important word, or words that you see onscreen. For instance, words in menus or dialog boxes appear in **bold**. Here is an example: "The **zBoot** option can be found on the second page of the **Instance Management** category, which can be accessed from the left-hand menu in **All automations | Service catalog | Instance Management**."

> **Tips or important notes**
> Appear like this.

Get in touch

Feedback from our readers is always welcome.

General feedback: If you have questions about any aspect of this book, email us at `customercare@packtpub.com` and mention the book title in the subject of your message.

Errata: Although we have taken every care to ensure the accuracy of our content, mistakes do happen. If you have found a mistake in this book, we would be grateful if you would report this to us. Please visit `www.packtpub.com/support/errata` and fill in the form.

Piracy: If you come across any illegal copies of our works in any form on the internet, we would be grateful if you would provide us with the location address or website name. Please contact us at `copyright@packt.com` with a link to the material.

If you are interested in becoming an author: If there is a topic that you have expertise in and you are interested in either writing or contributing to a book, please visit `authors.packtpub.com`.

Share Your Thoughts

Once you've read *ServiceNow for Architects and Project Leaders*, we'd love to hear your thoughts! Scan the QR code below to go straight to the Amazon review page for this book and share your feedback.

https://packt.link/r/1803245298

Your review is important to us and the tech community and will help us make sure we're delivering excellent quality content.

Download a free PDF copy of this book

Thanks for purchasing this book!

Do you like to read on the go but are unable to carry your print books everywhere?

Is your eBook purchase not compatible with the device of your choice?

Don't worry, now with every Packt book you get a DRM-free PDF version of that book at no cost.

Read anywhere, any place, on any device. Search, copy, and paste code from your favorite technical books directly into your application.

The perks don't stop there, you can get exclusive access to discounts, newsletters, and great free content in your inbox daily

Follow these simple steps to get the benefits:

1.   Scan the QR code or visit the link below

https://packt.link/free-ebook/978-1-80324-529-4

2.   Submit your proof of purchase

3.   That's it! We'll send your free PDF and other benefits to your email directly

Part 1 – Pursuit of Value

ServiceNow implementations, as with most worthwhile endeavors, are costly and difficult; if not for the substantial value they unlock, few companies would undertake them. This section discusses the strategy and tactics of using your ServiceNow implementation as a tool to drive business value.

This part of the book comprises the following chapters:

- *Chapter 1, Understanding the Value of ServiceNow*
- *Chapter 2, Recognizing and Avoiding Value Traps*
- *Chapter 3, Capturing Value from Your Implementation*
- *Chapter 4, Planning an Implementation Program for Success*

1
Understanding the Value of ServiceNow

We will start this book by discussing the value that can be enabled by the deployment of ServiceNow. All too often, the ServiceNow platform (and other tools) is deployed to fill a particular role within the technology landscape without focusing on the value it generates for the organization. When an organization sets out to implement a *ticketing tool* or an *HR portal*, it can be an indication that they're on this path. If you're reading this book, you have likely been given a key role in bringing ServiceNow to an organization and you would like to help make that implementation as valuable as possible. This book is going to give you concrete and actionable advice on how to achieve that goal.

This chapter will provide specific guidance on how to discover and shape the value proposition of your organization's investment in ServiceNow.

In this chapter, we're going to cover the following topics related to Managing Value:

- What is Value?
- Why is managing ServiceNow's value important?
- Who is responsible for value?
- How do we define value?

Then, we will dive deeper into four categories of ServiceNow implementation value:

- Service quality
- Cost optimization
- Customer experience
- Innovation enablement

Managing for value

When setting out on an implementation, your chances of realizing value will be much higher if you actively manage specific value outcomes. This section will provide an overview of value and its management in the context of a ServiceNow implementation so that you can clearly define the value expected in your own implementation.

What is value?

To achieve value, it is first necessary to understand what it is. Simply put, value is the benefit that your organization receives from some activity or investment. The value of a financial investment includes the price you receive when you sell it and the value of a piece of fine art includes the pleasure of owning and displaying it. Similarly, the value of ServiceNow includes all the benefits that result from its implementation. Value can be financial or non-financial, and it can accrue to yourself, your customers, or your shareholders. It is also sometimes referred to as the **return on investment** (**ROI**), benefits, or positive outcomes – in all cases, value is the desirable result of an investment of resources in the execution of some activity.

Why is managing ServiceNow's value important?

This book is predicated on the assumption that the leaders who sponsor and commission implementation projects such as a ServiceNow deployment are doing it because they wish to reap measurable benefits from the outcomes of that implementation. While some small minority of projects may be commissioned for other reasons (that is, political maneuvering), it's much more likely that someone is investing and hoping to see tangible benefits.

The value of ServiceNow must be clearly understood at the outset of a project to ensure it will be realized. If the planned benefits are not realized, then the implementation will have failed even if there is a go-live where cake is served (it is a tradition in the ServiceNow community that each major go-live is celebrated with a branded cake). While some would hope that simply introducing ServiceNow immediately provides you with a vast array of benefits, it is the careful and considered application of that tool to your company's business processes that will help you achieve the objectives. Actively managing the value of your implementation will result in a higher ROI, which, depending on your role, will translate to a more satisfied boss, customer, or stakeholder group.

Who is responsible for value?

The leader commissioning and sponsoring the ServiceNow project will ultimately be responsible for the ROI – for this reason, the final responsibility lies with the project sponsor. However, in many cases, this responsibility is delegated to a project leader who will chair the steering committee and serve as the person charged with the success of the project.

This project leader may delegate the responsibility for daily management of project work and outcomes further to a project manager. However, at the end of the day, a team will be charged to deliver the project's outcomes and the daily work of implementing ServiceNow will fall on their shoulders. What is essential, regardless of the number of levels in your project's governance structure, is that clear and continuous communication around value occurs from those doing the work to those responsible for the outcomes. The facilitation of this communication is the responsibility of the project leadership team and the project leader.

Alongside the primary responsibility of the project leader, there is the opportunity for every member of the team to take ownership of their contribution to the project's value. This ownership includes identifying value opportunities, documenting value statements, aligning the project execution with the desired outcomes, and drawing attention to any risks to value realization. In this sense, if you are reading this book, it is likely that you are responsible for value in your role in ServiceNow implementation and that by focusing on bringing success to your implementation rather than the often-stated goal of *going live, y*ou can make an outsized impact on the benefits that will be achieved.

How do we define value?

Defining value is difficult and many projects get launched and even completed without this essential step – however, if the value is not defined, there is little chance of it being realized. A useful practice for defining value is the creation of value statements that can precisely anchor the discussion of value:

- You should always describe the mechanism for value generation in your value statement so that it can be used to steer decision-making on your project.

- You should always define a value statement regarding some group or organization – even a brick of gold is only valuable if someone can make use of it and gold ore deep below ground is only potentially valuable.

- You should always strive to quantify value in a value statement, even if it's a non-financial metric – otherwise, you may not be able to determine whether a particular element of value was realized. At the end of the day, if you can't tell whether the value was achieved, then it probably wasn't.

Here's an example of a value statement - **Automate widget request sourcing**: The introduction of automation into the widget request process at the sourcing step reduces effort on the part of the widget manager by 15 minutes per request, with a cumulative annual savings of 460 hours, while reducing the time customers wait for their widgets by 1 day.

In almost all cases, value statements should be documented in a location that is accessible to your project teams, except for value statements relating to confidential activities such as unannounced organizational change or a material event such as an acquisition. By making the value statements available to your project team and referencing them throughout your project, you increase the probability of these benefits being realized. In projects being delivered by your own organization, your project charter or project business case may incorporate the value statements. In the case that a consulting or professional services organization is working on your implementation, the documentation on the project scope of the work can serve as a useful place to define these outcomes and to ensure alignment between all parties.

Types of value

While the value proposition of each project in each organization may be slightly different, years of implementing ServiceNow across diverse industries have shown that some common themes re-occur. Each of these provides a significant opportunity that should be explored in the context of your implementations, but it is most common for organizations to choose one or two as a primary focus of their projects. The four general themes are the following:

- Service quality
- Cost optimization
- Customer experience
- Innovation enablement

Each theme deserves independent treatment and the remainder of this chapter will be dedicated to describing these areas of value and the specific considerations that should be applied to the ones you target. In particular, we will review the important information needed to effectively drive value for each area to support you in determining which value opportunities you will target and what activities, features, or modules you will include in your scope to achieve those opportunities.

Service quality

Perhaps the most classic ServiceNow use case, the theme of *service quality* relates to an organization's ability to deliver what they have committed to. The nature of services may vary but the underlying principles are quite similar and even the terminology used to manage them is often aligned. In an IT organization, service quality is the availability and smooth operation of the enterprise software and hardware, from laptops to Wi-Fi and from email to accounting systems. In a telecommunications company, service quality is about the stability and speed of communications services such as the internet and cable TV, while for an internet company, the availability of their website might be the primary consideration.

Defining a service scope

When striving toward service quality improvements, it is critical to define a *service scope* consisting of the services whose stability we aim to improve or maintain. This is easiest in an IT organization where the service scope might be "*all systems managed by IT*" or more commonly, "*All production systems managed by IT*," but again, the same principles can be applied to other domains.

ServiceNow supports service quality primarily within the **IT Service Management** (**ITSM**), **IT Operations Management** (**ITOM**), and **Customer Service Management** (**CSM**) product areas, although other products are involved for particular use cases as well. The starting point for most organizations focused on service quality improvement is usually the establishment of ITSM processes for in-scope services. In these cases, it's common to see scopes targeting *all data center devices*, or *everything on the network* – the underlying assumption being that if all the individual components are running smoothly, then the collective service is also performing as expected.

In defining a service scope, you may include components such as servers, switches, or storage arrays, and you will also need the list of business or application services that you are targeting. With ServiceNow implementation in particular, it is essential to at least define the applications that these components support, as these form one of the cornerstones of the ServiceNow **Common Services Data Model** (**CSDM**) as *Application Services* (more detail on the CSDM can be found in the current version of the ServiceNow published "*CSDM Whitepaper*" available from ServiceNow).

Some guidance on service scope is warranted, as companies often try to tackle all systems and services for service quality optimization simultaneously – while that is possible, it also requires an expert level of coordination and is subject to significant risk. A more targeted approach can remove risk and decouple these objectives, allowing you to move rapidly in a small domain to deliver value quickly. It also allows an implementation team to build confidence by tackling a smaller problem with a clear end in sight, rather than a single goal that is difficult to envision reaching and even more difficult to execute.

Service quality metrics

It is common for projects to be commissioned with high-level value objectives tied to a value theme such as *service quality*. Organizations may know that there is work to be done to keep things running smoothly but have not yet translated that into actionable metrics. In these cases, the project team will now need to collaborate with stakeholders such as business and IT leaders and service owners to create clear value statements for service quality tied to specific service quality metrics. Some examples of value metrics related to service quality include the following:

- The number and severity of SLA or SLA breaches
- The number of critical incidents
- Unplanned downtime of mission-critical systems
- Failed customer interactions with your mobile application

Note that while these are quantitative metrics that undeniably matter to your organization, they may or may not be financial metrics. An SLA with your customer may incur penalties for being violated but an internal OLA isn't likely to have a hard dollar cost associated. This will be a common trend in your ServiceNow journey and will be repeated for other value-related themes. Recall that a value statement should be measurable in terms of the impact on a specific group of individuals. This requirement should be easy to meet when formulating service quality metrics and value statements, as you generally know what the user demographic of a service is. Examples of user communities can include employees in *the accounting department*, your *customers*, or your *suppliers*.

A value statement for service quality might read as follows - **Reduce the impact of database anomalies during the month-end close**: Monitoring and responding to anomalous database performance during peak periods around the month-end close reduces overtime worked by the financial accounting team by 50%.

This value statement is clear in scope and delivers value to a clear audience, although whether the accountants are working long hours or the budget holder is footing the bill for the overtime pay may depend on the organization. By targeting specific systems or services that have a measurable impact on the organization, you avoid a peanut butter approach where a diluted version of the value is spread across a large surface area of applications. This is important because a very small impact on a large number of teams can be difficult to measure and it therefore becomes difficult to show that an improvement is related to the efforts of your project team.

Planning for service quality

With the metrics established, the next task for your team will be to develop and execute a sequence of activities that lead to the realization of value. A critical step will be to confirm that you have the appropriate ServiceNow licensing in place to support the service quality value objectives and that your project work plan includes the activities required to deliver on those value objectives.

If there are any gaps, you must work with your project management and governance structure to either modify the plan or the objectives. This is an essential point, as all too often, the connection between objectives and the plan is ignored. If you as an insider within the project, a dedicated member of the team, or a seasoned leader overseeing the project can't tell how a particular objective will be achieved, then there is a good chance that it will not be.

Identifying opportunities in the current state

If service quality is a target outcome for your project, then you should understand the current state of the in-scope services first. This is a common challenge where organizations believe that the current state is *not* relevant to their future state. They are re-imagining their IT operations and implementing an entirely new ServiceNow instance after all. Experience shows that a good map is substantially more valuable if you know where you are starting and for that reason, it is worth your while to get a sense of how the processes are currently performing.

Questions to ask include the following:

- How stable are the services?

- When something goes wrong, how quickly is the issue detected?

- When investigated, what are the common root causes?

- How often are issues resolved at the service desk versus being resolved only after escalation?

Answering these questions will help you map specific ServiceNow capabilities to the problems most relevant to these services. This, in turn, allows you to focus your resources on the key levers to drive improvements in your objectives.

Aligning the implementation scope with the opportunity

If you find that the in-scope services often fail due to changes being made to those services, then that suggests it's necessary to focus on the change management process to better assess and manage risk. ServiceNow has specific capabilities in ITSM for change risk calculation to help you navigate that scenario. However, if an unreliable software solution, perhaps one prone to memory leaks, is spontaneously failing after some time in stable operation, then the instrumentation of that infrastructure and application using the ITOM monitoring capabilities should be considered to identify leading indicators and even execute automated remediation steps as a stopgap until the software is updated.

The essential process of discovering the problems or opportunities that exist in the service quality requires you to speak to those who are knowledgeable about the services, analyze the data, and think critically about how the capabilities you will implement can bring about a positive change for their teams. During this time, keep an eye out for teams who have substantially better service performance or who have established and proven solutions for managing their systems.

If you are operating from a templated plan or even one devised at the outset of the project when you had very little information, now is a good time to revisit your planned activities and to critically assess whether changes are required to remain on track to deliver value.

Cost optimization

When constructing a business case for any project, a common activity is to identify how the implementation will provide returns on your initial investment. If you have consulted this business case when preparing your value targets for cost optimization, you will have a head start in terms of the direction of your project's documented objectives. Some of these objectives may not become part of the final value proposition for your project but they do set a direction for focus and a baseline for comparison.

Cost optimization opportunities in ServiceNow implementations often target one or both of the following major drivers of economic value:

- **Process efficiency**: Reduction of the time spent executing and waiting for common tasks to complete
- **Asset optimization**: Efficiency utilization and stewardship of company assets

We'll address the opportunities in these drivers of value so that you can better understand the options to realize the financial ROI of your implementation.

Process efficiency

This is the classic ServiceNow business case metric, the reasoning being that if we spend thousands of hours handling thousands of tickets every month, then even a small improvement quickly pays for the cost of the implementation. Usually, some percentage ranging from 3 to 15% is selected as the target efficiency gain, but at the planning stage, it is rarely clear which specific activities are the targets of these gains. At the same time, some areas of value that are often overlooked in this analysis include the benefits that accrue to those customers whose wait time for a service is reduced or where reducing escalation and follow-ups can avoid taking time away from their daily work.

Understanding the current state

When embarking on a process optimization journey, you will be more well equipped if you can observe as many of the relevant processes as possible in their current form. A week of shadowing ServiceDesk or customer support agents, analyzing current ticketing data or inboxes, or other views into the current state will be time well spent, as you will gain insight into the way work is currently happening and both the benefits and drawbacks of the current processes. In many cases, even where tooling is not in place to support processes, the teams responsible for getting the work done will implement ingenious workarounds to make their daily lives more efficient but might not think to mention them in a formal requirements workshop. Even if the current processes are in no way better than the target state, you can still learn much about the degree of organizational change that will be required to shift to new ways of working.

Sequencing investment in processes

If there are specific processes that have been targeted for optimization in the charter of your project, then you can begin with those processes until clear reasons emerge to modify that scope. In most implementations, however, there is a general desire to introduce process efficiency and it falls to the project team and process owners to determine the sequence of implementation.

Again, in this domain, we risk spreading a thin layer of value rather than concentrating efforts to make an impact that justifies the further investment, so it is important to be intentional in your distribution of efforts. It is likely that simply by deploying a modern, fully integrated service management capability, your ability to serve customers will increase slightly across the board, but the opportunity for optimization doesn't stop there. The highest returns come when you tackle specific high-value requests or workflows to automate delivery and minimize cycle times to the tune of days rather than minutes. We will now investigate how to identify those opportunities.

Opportunities to optimize processes can be identified using the following sequence of steps:

1. Start by generating or collecting a list of request types across the teams and services in the scope of your implementation. These may be requests for goods or services, inquiries, or any other interactions that consume the time of the operational teams.

2. Next, score each of these opportunities by the monthly effort invested (the count of requests multiplied by the effort per request) as well as the potential savings from automation. In this context, automation not only includes fully scripted process executions but also the routing and batching of work to ensure that tasks are distributed efficiently and do not get lost mid-workflow.

3. Now, assign priority to tasks that score highly in both categories. In this way, a focused effort on optimizing even a small number of high-impact processes can translate to an outsized impact on your organization's total efficiency.

4. Finally, create value statements for the top priority items in order to ensure that there is a clear objective for each opportunity.

Again, processes specifically mandated in the charter of your project should be included on this list by default unless a strong argument is made to and accepted by your steering committee. In general, the approach we advocate is to preserve and meet commitments that have been made by the project when doing so does not risk loss of value. This establishes credibility that is useful down the road when a significant change is required in order to preserve value.

Leaving a process off the priority value list at this stage could mean that the process is no longer required and can be left out entirely – however, it is also possible that a basic form of that process can be implemented rapidly, allowing the team to focus efforts on the items that generate the most value. In the latter case, in the meantime, the basic implementations can capture volumetric data that will inform the next prioritization cycle.

Identifying improvement opportunities

A common approach in projects targeting process optimization is for a member of the team to sit down with a process SME and essentially generate a flowchart of the steps that must be executed in the delivery of that process or service. This activity is a useful information-gathering exercise, but an opportunity is missed if this flowchart is then handed to a development team who are instructed to build it into a workflow without engaging in process re-design. This is an intensive exercise that should be completed only for the processes identified on your prioritized list, specifically the processes your project has committed to improving and those that will generate the greatest savings for your organization.

When optimizing a process there are several tools ServiceNow provides:

- **The ability to require information at earlier stages and present it later in a workflow**: This is very important when you observe processes that frequently revert to earlier stages due to incomplete information.

- **The ability to validate the information in real time in order to correct errors before they impact the process**: This can be as simple as using a lookup to the CMDB to ensure an entered serial number corresponds to an actual device, or it can involve more complex validations referencing customer entitlements.

- **The ability to look up related information for the user rather than requiring manual entry**: This saves the user time when filling out the forms and prevents gaps in the user's knowledge from impacting the process.

- **Automation of routine actions**: Activities that can be completed automatically, such as the provisioning of cloud resources according to a template or the allocation of an unused software license from a pool. Automation is particularly valuable in conjunction with automated lookups – this combination reduces effort of fulfillment and allows for reduced customer wait time, even allowing for near-instant fulfillment of some requests.

- **Parallel execution of workflow**: The ability to route parallel execution of activities while managing prerequisites is one of the useful features of SerivceNow's Flow Designer and legacy Workflow tooling. This can prevent unnecessary wait times for the end customer, as parallelization allows processes to complete much earlier.

- **Process monitoring controls**: ServiceNow allows for due dates and SLAs to be configured so that overdue or stalled tasks can be identified quickly and attention can be given to completing them. Stalled or unattended tasks can often be identified in current state ticket data and are easy opportunities for optimization.

Taken together, a combination of these methods will help you drive down the effort and time taken to complete the high-volume processes identified earlier, which can represent a substantial saving in time and effort.

Here's an example of a value statement for process optimization - **Automate license allocation and software provisioning**: A catalog of department-specific pre-approved software that is requested for immediate deployment using existing software licenses enables users to get right to work with the tools they need by reducing provisioning from 24 hours to 5 to 30 minutes depending on the application.

Note how this re-imagined process references a user's department and looks up approved software to cut out unnecessary approvals by allowing managers to simply pre-approve certain software for all their employees. It also avoids inconvenient delegation processes when a manager is out of the office or effort for every new employee. It also refers to the automation of license checks, a key opportunity that is discussed in an upcoming section on *asset optimization*.

Process efficiency metrics

Process efficiency metrics are one of the areas ServiceNow is most optimized to collect data for due to the years of work in monitoring the execution of ITSM processes. Specifically, the Performance Analytics capability will allow you to collect detailed breakdowns of your process according to key dimensions and will allow you to monitor process performance to ensure it remains within specific bounds.

Common process efficiency metrics will include the following:

- **Process cycle time**: The end-to-end delivery time for a process
- **Time to live**: How long it takes from the request of equipment until that equipment is actively up and running
- **Stale task counts**: How many tasks have remained in a particular state without updating for longer than expected
- **Process throughput**: How many requests are being processed
- **Effort per request**: How many hours of hands-on work are required in order to serve a single request

Ideally, all these metrics would improve immediately at go-live, but it is often the case that the implementation of new processes and systems introduces a learning curve that initially offsets gains in efficiency. The effect of this is that things temporarily get more difficult as staff familiarize themselves with the new system before the benefits of the new processes start to provide a net benefit.

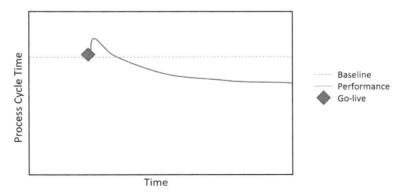

Figure 1.1 – An example of a process learning curve

Figure 1.1 shows a small but noticeable increase in the process cycle time at go-live, followed by a steady decrease that settles below the baseline cycle time as process participants become accustomed to the improved process.

Asset optimization

One of the most overlooked areas of value in ServiceNow is the ability to create a central register of company technology assets and define the processes and controls for their use. This can reduce the number of expensive compute devices purchased, extend the useful life of assets, and help you manage risks when decommissioning these assets.

Prerequisites

Asset optimization in ServiceNow is primarily achieved through the Hardware Asset Management (HAM) and **Software Asset Management** (**SAM**) modules, although supporting ITOM and ITSM capabilities often come into play. These are some of the modules with the greatest potential to optimize technology expenditure but are generally under-used in the ecosystem.

If your organization is focused on asset optimization as an outcome of your project, then ensuring you're planning for the implementation of these capabilities is essential. Asset optimization processes are also very dependent on data quality and availability – often, interfaces with vendors, procurement systems, and financial systems are required, so you should be sure to include these in your project scope. SAM is also dependent on the collection of data from your in-scope devices in order to determine where various software is installed.

The definition of an asset

A common question during implementation in terms of an asset optimization component is the definition of an asset and while there may be any number of theoretically correct answers, rather than asking *what an asset is*, you may instead consider the more applicable question, "*when should we create an asset record in ServiceNow?*" The former question is theoretical – the latter is a question of system behavior and what capabilities you'll have, so it's more immediately useful.

An asset record should generally be created if any of the following three conditions are met:

- The item holds financial value or financial liability is associated with it
- The stock levels of the item must be maintained, either as an individual item or in bulk
- The process of procuring, deploying, and decommissioning the asset should be tracked across its life cycle

The focus of the asset record is to identify, classify, locate, and value the item it refers to while the associated CI record will address its operational characteristics.

Hardware Asset Management

When organizations purchase hardware, it can quickly become difficult to keep track of it. In fact, in many cases, items worth less than a couple of hundred dollars are considered *consumables* and distributed as needed but without a formal process to manage their whereabouts or life cycle. An extreme example of this would be printer paper. It's not important to know who has which pages but it is important that we know when we're running low and need to re-order. ServiceNow provides distinct processes and options for managing consumables relative to individual serialized assets such as servers and laptops.

Determining what assets should fall into each type requires you to consider the value of the assets, the rate at which they are used, and any relevant accounting or compliance requirements; furthermore, some organizations lease their assets and as a result, individual tracking is essential in order to ensure that assets are returned at the end of the lease period.

Software Asset Management

Software licenses come in many different types and ServiceNow has an extensive capability to manage these considerations. The basic operation of SAM is to ensure compliance by determining whether the company has enough licenses purchased or subscribed to for the actual usage. This is a pre-emptive self-audit and helps companies avoid expensive true-ups and penalties, which vendors can apply if their customers are found to violate their licenses.

When considering ServiceNow for SAM, you should also determine whether an existing SAM process or tool is in place and consider the incremental value ServiceNow brings. If another tool is deployed, is it efficiently handling the interfaces to service management (for example, reclaiming desktop software licenses when a laptop is returned)? If a manual process is involved, can time be saved in the execution of those processes that can be redeployed to support contract negotiations?

Scope for asset optimization

As with service quality, the starting point for asset-based cost optimization is understanding your scope. The broadest useful questions for scope are the following:

- Does your scope include both software license and hardware assets?
- Will you manage both serialized (individually numbered and tracked) as well as consumable assets?
- Are you targeting both data center and end user assets?

These questions determine your high-level scope and narrow down the discussions to follow. Once the high-level scope is defined, you will need to establish boundaries for the process, asset types, and organizational scope in order to focus your implementation. Naturally, you will want to consider an objective when setting the scope and prioritizing the scope that offers the greatest potential for cost optimization.

Useful scope groups include the following:

- **Process**: Are you trying to implement an end-to-end life cycle, just the procurement and provisioning, or just the assignment and tracking? ServiceNow provides tools to do any or all of these but it's up to the customer to decide which of the processes should be prioritized.

- **Asset types**: The specific types of assets that should be included should be defined at a level more granular than the data center or end user. You should determine specific categories to target, such as rack-mounted switches, servers, laptops, printers, and wireless access points.

- **Organizational scope**: Will you just target your corporate offices, your field staff, or every employee across certain geographies? Another common question is whether to include new acquisitions either as a pilot group or for later rollout.

There are no wrong answers when it comes to the scope of assets to include, but if you choose to attempt everything at once, you are increasing the complexity of and therefore the risks associated with your implementation. We recommend choosing at most one of the aforementioned scope areas to cover broadly in your first release. The greater the skill and expertise of your implementation team, the larger the scope can be for a given release.

Sample of a focused asset scope

To establish procurement and provisioning processes for all desktops, laptops, and laptop docking stations in corporate offices worldwide.

Asset optimization metrics

Measuring asset optimization metrics can be more challenging than process optimization but is still well supported by ServiceNow.

Key metrics of asset optimization include the following:

- Spend per headcount on end user devices

- The annual asset loss rate

- Asset utilization rates

- Any software asset license penalties paid

- Software assets re-allocated

- An inventory of critical spares

- Embodied carbon of corporate devices

- Asset energy utilization and emissions

In addition to the actual value metrics, you should also consider data quality metrics such as the following:

- Stale asset records

- Duplicate asset counts

- Assets with invalid location, financial, or assignment data

- Missing asset aging

These metrics are just as important as your optimization objectives because they will inform the trust you can place in your other metrics. In many cases, organizations will require periodic audits of assets for operational or accounting reasons, which can give you an excellent opportunity to sample these metrics.

Customer experience

ServiceNow may have grown from humble beginnings as a platform to support IT processes but today, it offers a substantial capability to improve the experience of customers, whether those are internal or external to the organization. In our modern lives, we often take the consumer-grade experiences we encounter with apps such as Uber, Amazon, and Netflix for granted. ServiceNow can enable a version of this consumerized experience to our stakeholders, whether those are end customers, employees, supply chain partners, or job candidates.

The **HR Service Delivery** (**HRSD**) application provides the ability to onboard new employees seamlessly, meaning that their first experience in your organization is far less overwhelming and confusing than it might otherwise be.

The CSM application allows you to provide contextual, entitlement-driven support at scale to your customer base while supporting the integration of omnichannel engagement with those same customers.

In ITSM and enterprise portals, in general, end users can *shop* online for the services and equipment required to get their jobs done and take advantage of published knowledge bases that provide relevant information at their fingertips. This is useful to employees in major offices, but you could also consider retail or branch locations as well as field employees and their needs.

Customer experience is a value proposition that is usually either scoped directly into your project in the case of a CSM or HRSD implementation or it may be a primary or secondary consideration in the deployment of other ServiceNow modules such as Legal Service Management or ITSM. This means that for many implementations, the scope of customer experience is clearer than some of the other value drivers, but what can be particularly difficult is how to measure it. The tools remain the same regardless and, in this section, we will focus on the things you can do in your project in order to maximize the customer experience.

Tools for customer experience

The most important part of customer experience is whether your customers are consistently and efficiently receiving a service that meets their needs. For this reason, many of the elements covered earlier in this chapter apply, particularly process efficiency and service quality. In addition to stable services and quick turnaround times, your customers and stakeholders have grown used to the experience of being able to digitally interact with services on their terms and in their own time. This can mean self-service portals that put knowledge and structured workflows at the customer's fingertips, as well as Virtual Agent chatbot conversations that can automatically handle complex interactions and hand them over to a human when necessary.

There are two major items that both drive customer experience value and quickly expand the scope of your projects: portals and Virtual Agent conversations.

Scoping customer experience value – portals

The portal scope is often defined at the outset of the project as simply being out of the box with the assumption that this will meet the organization's needs. This may be the appropriate course of action but if you are planning to use an out-of-the-box ServiceNow portal for your deployment, you should be certain to conduct a detailed walkthrough with representative users in order to determine whether it will meet their needs.

Fortunately, branding a ServiceNow portal is a straightforward process and aligning it with your company's image is essential if the portal will serve as an interface outside your organization. For more extensive customization, you will need to ensure there is a clear value statement tied to each departure from the ServiceNow default in order to ensure that the changes are being made for the right reasons; but if justified, ServiceNow portal technology is flexible enough to accommodate just about any set of requirements.

Scoping customer experience value – Virtual Agent

One of the innovative capabilities of the ServiceNow platform is the conversational *Virtual Agent* system that allows you to combine natural language processing and an automated workflow to generate self-service options that can be as good as a live agent but at the disposal of your users on their schedule and without your support staff's involvement. Virtual Agent conversations can be prioritized and scoped in a similar way to investments in process efficiency workflows from earlier in this chapter. By creating a combined view of interaction frequency and then assessing the degree to which the interaction can be fully scripted, you will be able to identify the best opportunities for Virtual Agent topic delivery.

A useful exercise with Virtual Agent conversations is to analyze logs of any existing chat-based tooling for support and use this as a basis to prioritize automated conversations. It is also important to realize that if your virtual agent is not able to address a large fraction of incoming requests acceptably (even if it means transferring to a live chat agent), then the agent is more likely to frustrate users and lower the overall customer experience. Consider designing your flows so that if a clear actionable intention is not quickly established and confirmed, then the virtual agent gets out of the way and lets a human take over.

Measuring customer experience

Measuring customer satisfaction with the experience you provide is a challenge because traditional methods such as surveys can have low response rates. For this reason, it is important to engage early and often with representative customers in order to get direct feedback and make good use of the collected data and feedback.

ServiceNow provides survey and rating capabilities that can be used to receive feedback from customers. You must resist the temptation to create surveys that are time-consuming or that contain many required fields. Often a simple happy, neutral, or sad face with an optional text box for explanation can capture more useful data than a five-item survey because you'll get substantially higher response rates.

Innovation enablement

The final area of value we will cover in this chapter is the enablement of your organization's ability to innovate and execute new opportunities. This is typically not something tackled as a primary objective in an initial implementation of ServiceNow but there is a lot of value to be unlocked here and so we will cover the opportunities and approaches to enable innovation in an organization.

In this section, we'll cover a diverse set of innovation enablement capabilities that are worth discussing for their ability to either support the formulation or prioritization of ideas within your organization as far as they strengthen the ability to execute projects and products.

Ideation

Included in the ITBM product family is the *Idea Portal*. This is a tool that helps you efficiently source content from your organization and assess the potential of executing those projects. Even if the Idea Portal is not a good fit, the concept of centralizing the intake of helpful suggestions or ideas through demand management is something that can be incorporated into almost any service catalog. It also prevents the contamination of incident and service request data with a very different type of interaction where service stakeholders are suggesting ways to make a change to the service rather than interact with it operationally.

The value of implementing the Idea Portal or a demand management intake channel early in your implementation journey is that you can use it to help gather information relevant to your implementation as well. Similarly, the introduction of the demand process early in your implementation journey enables you to tie into the project management capabilities, which can help drive the execution of your ServiceNow implementation and other enterprise projects.

The management of ideas and demands therefore provides both direct benefits in the aggregation of knowledge within your organization and improvements to data quality in other processes that might otherwise serve as a channel for demand intake in the absence of a more suitable path.

Project management

Another significant challenge in modern organizations is a lack of standardization in the execution of projects, in large part due to the lack of widely adopted collaborative project delivery tooling. If your organization is still tracking risks and issues in Excel or PowerPoint, then you might consider the introduction of ServiceNow's project management capability. One very effective way to pilot this is by adopting it for your own ServiceNow implementation, which has the added benefit of getting project participants more familiar with the tool and its features from an end user's point of view.

Specific areas of value that can be piloted on your implementation before releasing it to the broader organization include the following:

- **Online fully collaborative project planning and execution**: Rather than a single project file on a manager's computer, you can grant project participants a real-time capability to update their parts of the plan and review the progress of those they are dependent on.

- **Transparency in risks, issues, decisions, actions, and changes**: The online **Risk, Issue, Decision, Action, Change** (**RIDAC**) capabilities within ServiceNow **Project Portfolio Management** (**PPM**) help surface key project management risks and issues and provide a standard workflow for addressing them. They also serve to enable greater accountability, as all actions can be tracked and their status can be reported.

- **Project status reporting**: ServiceNow allows you to generate in-system status reports and share them with the right stakeholders. Relevant data on project progress and cost can be included to present a clear picture of progress.

In addition to facilitating project delivery, ServiceNow also offers a significant opportunity to accelerate technical execution through the inclusion of agile and test management capabilities directly on the platform. These features integrate into project planning as well, which facilitates even greater transparency and reduces the potential for missed communication.

Summary

This chapter has covered a lot of ground, from addressing the foundational concepts in a value-driven implementation to walking through four broad categories of value in ServiceNow implementations with specific examples. Using the information in this chapter, you can now articulate the need for active management of project value in order to maximize the ROI in ServiceNow. Additionally, you should understand the broad categories of value within a ServiceNow implementation and be able to use this to frame and document specific value objectives for your project. Finally, you should be able to connect the value that you are targeting to specific features and capabilities that will form your project scope and specify metrics that will track the success of your value objectives.

In the next chapter, we will look at some of the common ways in which projects fail to deliver the anticipated value.

2
Recognizing and Avoiding Value Traps

In the previous chapter, we talked about the different ways your implementation can bring value to your organization. Now, we will address some of the common pitfalls that impact ServiceNow implementations and prevent the realization of the targeted value.

In this chapter, we will cover five value traps that re-occur across implementations. By steering clear of these issues, you will be far more likely to achieve your project's objectives. If you're a member of the team, this chapter will also help you articulate the risk of these value traps to your project leadership and argue more convincingly for an approach that avoids these common pitfalls. The value traps covered in this chapter are the following:

- Replicating the current state
- Ignoring the current state
- Chasing the long tail
- Not managing change
- The science experiment

These value traps are common because, at first glance, the courses of action that bring them about are attractive to project teams and leaders. The result of these approaches, however, is consistently problematic, as it tends to impair your ability to deliver certain types of value.

For each trap, we'll cover the approaches that lead to the value trap, the reasons why that approach is attractive, and the issues that it can cause. Finally, we'll address alternative strategies and considerations that help provide you with options to balance these issues beneficially.

Replicating the current state

The value trap of *replicating the current state* occurs when an implementation defines a less-than-ideal process for deployment because that is the way it has always been done in the past or the previous system. The way things are done now is often called the *current state*, while the future design is the *target state*. Often, making the target state match the current state leads to customizing the ServiceNow platform away from the **out-of-the-box** (**OOTB**) processes and towards an outdated, inefficient, or unmaintainable configuration that will complicate upgrades and lower your Instance Scan scores.

The tool replacement approach

The replication of the current state, also known as a like-for-like replacement, is most often encountered when the project is seeking the shortest path to go live with the new solution. These projects often have the goal of replacing some existing system, process, or technology and a time constraint under which they must execute the transition. Most people associate tool replacements with the decommissioning of a competing system – for example, in ServiceNow's case, you might find BMC Remedy, HP Service Manager, or something similar as the incumbent technology. However, in addition to these cases, replacing an Excel file, SharePoint list, Access database, or even paper-based forms can lead to a tool replacement methodology. Even the replacement of a legacy ServiceNow instance can start to take on characteristics of a tool replacement.

Some elements of your implementation will likely target the replication of the current state, while others will avoid this approach entirely. As with other value traps, it may impact some, or all, of your project's scope.

Arguments in favor of replicating the current state

Replicating the current state would not be a value trap worth mentioning if it wasn't frequently encountered in implementation projects and it wouldn't be as common as it is if not for the fact that there are many potential or perceived advantages. In some cases, your project has been commissioned to deploy a more modern solution that avoids the costly maintenance of legacy systems or to address unstable and failing applications – the saying "*time is money*" applies quite literally as the costs to maintain existing technology grow. In other cases, there may be a great deal of institutional knowledge embedded in processes that have been developed over years or even decades. The idea of altering these processes can seem to be an insurmountable amount of work that falls outside the budget for the implementation. The users of the old systems may also feel very comfortable with their processes and resist any changes to the way things have always been done.

It's important to recognize some of the key benefits of a tool replacement methodology to understand why so many team leaders are attracted to it:

- A tool replacement (also negatively labeled a *tool slam*) can appear to have far less process design required because you are simply adopting known and defined processes

- Replicating the current state creates a similarity between the old and the new processes, which also implies a greatly reduced effort for the management of changes associated with the project

- Finally, the tool replacement project appears to have a very clear scope because you have a clear specification of functionality in the form of the current system or process

Issues with replicating the current state

We can see that there are good reasons why it is appealing to replicate the current state process but, as with so many decisions in an implementation, no path is without its advantages and disadvantages. In this section, you will learn about the common drawbacks to the replication of the current state, and it will then be up to you to balance these considerations against the arguments in favor of current state replication while clearly understanding the impact this will have on the value realized in your project and its long-term **return on investment (ROI)**. The following concerns are most pronounced when the legacy process is different from the ideal target state in ServiceNow, often because those processes have been tailored and built up over the years.

Replicating tailored current state processes can be problematic for the following reasons:

1. If your existing tools are overly customized, difficult to maintain, or unstable, the last thing you want is to bring a similar degree of complexity to the ServiceNow platform. A new implementation should be an opportunity to start fresh with a clean slate technically.

2. Your current state processes may also have nuances or customizations that have accrued over years of use that are not well documented. Frequently, processes evolve to handle dozens of edge cases, each of which needs to be considered and configured in ServiceNow. This additional effort in discovery, documentation, and development can quickly outweigh the savings of not designing a streamlined process.

3. Legacy processes often do not take advantage of new capabilities of the ServiceNow platform such as Virtual Agent, contextual search, and Predictive Intelligence. By replicating the capabilities of an older tool, you neglect the value that could be obtained by deploying some of the features that the current ServiceNow version supports.

4. The legacy processes have likely been implemented in a way suited to the architecture of your previous tools – an optimal process in ServiceNow might look very different if designed specifically with the ServiceNow capabilities or OOTB processes in mind.

5. The quality and consistency of data created using legacy processes may also be low, preventing you from using a data-driven approach to get to the root causes of issues and optimizing efficiency.

In general, the replication of the current state prevents you from realizing some of the value of investing in ServiceNow while presenting the risk that you will carry the technical debt that prompted the tool replacement forward into your new ServiceNow instance. The cost of time and effort of dealing with nuances of the old process flows often erodes the benefits that prompted the like-for-like strategy in the first place.

To present compelling alternatives to the current state replication strategy, we will review two strategic options that can provide many of the advantages while mitigating some of the issues.

Strategy 1 – Adopting the out-of-the-box process

The first alternative applies when the default ServiceNow platform has features that can facilitate the same outcome as your legacy process. In these cases, you should recommend the adoption of the OOTB ServiceNow process instead of replicating your current state. This approach has the advantage of accelerating the detailed process design of your implementation but, unlike the like-for-like scenario, it does not result in substantially increased technical efforts. This is because you can utilize ServiceNow best-practice processes and their associated configurations instead of replicating the old solution's features (which are often not available out of the box in ServiceNow).

This alternative has only one significant drawback relative to the like-for-like approach – training and change management activities are likely to consume more effort, as you'll have to transition teams from the old way of doing things to the newer model. Generally, the greater the difference between OOTB ServiceNow and your legacy process, the larger the scope of the change management effort that will be required. You will likely need to create documentation that maps the user's familiar current state into a target state process and monitor for cases where people revert to their old ways of working.

Fortunately, these change management and documentation efforts can be offset by the corresponding savings in development, testing, technical documentation, and ongoing support. These are likely to be larger and help balance the scope of process change.

An example of an area where this approach applies very well is in the Hardware Asset Management space, where the data model and workflows are tailored to an industry standard way of managing assets that goes above and beyond what most competing solutions offer.

The application of OOTB processes can also provide unexpected benefits due to the inherent synergies of ServiceNow's single platform approach. For example, incident management executed according to ServiceNow's guidelines will create information that can be leveraged for service level management and knowledge management and is automatically populated in standard workspaces and portals.

Strategy 2 – Developing an MVP process

The second alternative to a like-for-like approach is the development of a lightweight or **Minimal Viable Product** (**MVP**) process in ServiceNow that allows your users to interact with a stripped-down version of the process with the bells and whistles removed. This approach is often warranted in customer cases, IT service requests, or HR cases where OOTB flows may not cover the intended use case and ServiceNow provides the toolkit for the construction of arbitrary forms and flows.

When taking an MVP implementation approach, it is important to clearly outline the scope constraints that will be applied because, by definition, an MVP will not achieve all the goals you might hope for but rather, only absolutely necessary objectives. A documented scope will help you effectively manage changes to that scope with a clear picture of how it will impact your resources and timeline.

Targeting MVP processes can also provide a substantial reduction to the design and development effort, while the simplicity of these processes may also reduce the training overhead for your project. The trade-off is that if your project has value objectives related to process optimization, you may struggle to realize those with only a simplified process implementation.

> **Important note**
>
> In cases where the legacy process is simple and does not require a complex workflow or custom features, the MVP approach ends up being most similar to the like-for-like scenario. This is not a concern because issues 1, 2, and 4 (as we talked about in the *Issues with replicating the current state* section earlier) would not be impactful and your MVP processes can be uplifted later to leverage platform capabilities more efficiently.

Ignoring the current state

Closely related to the first value trap is another failure mode where instead of replicating the current state, the project leadership marks it as clearly out of bounds for analysis and charters the project team to only consider the organization's target state. Another form of this value trap occurs when analysis of the current state is considered separately from the target state design and change management plan by an independent team.

Focusing on the future

When developing a new set of processes and supporting capabilities to enact significant change to the way things are currently done (which is often the case when seeking service quality and cost optimization value), it can be important not only to avoid replicating the current state, as we discussed in the previous section, but also to re-imagine and re-engineer processes to improve them. Decision-makers with limited resources will need to prioritize efforts and may decide to only focus on the future rather than conducting shadowing, research, and data analysis to understand how things work today. This approach can be appealing for the following reasons:

- The current systems and processes are often being replaced because they are inadequate or suboptimal. Why would the team want to spend time studying these inferior solutions rather than designing and implementing a better set of processes?

- The current state can be seen as irrelevant to the future design because it is being replaced, decommissioned, and discarded. This argument focuses on the fact that elements that are being discarded during the implementation will not have an enduring effect on the future performance of your organization.

- There will be a risk that the old ways of working influencing the new way could impact the organization's ability to change. We're already discussed the many reasons to avoid replicating the current state and these are often used as a reason not to investigate those processes too deeply.

Failing to learn from the past

All three of these arguments suggest that we focus on spending time where it matters most – the future. What these arguments fail to capture is the importance of the journey of transitioning from the old to the new and the fact that the past can both inform the future, as well as help us understand exactly what will change during the implementation, which allows us to manage that change more actively. Opportunities that are missed in implementations that ignore the past typically include the following:

- Without investigating the current state, you will likely miss out on certain stakeholders for processes. If someone is involved in a process today and that process changes significantly, then that person or organization should at the least be considered as part of the change management. They may be making extra accommodation for gaps in the process or relying on report data in certain formats. Walking through the current state allows you to better identify these people to reduce friction in the **User Acceptance Test (UAT)** and go-live phases.

- One advantage the past processes have is that their performance is already well established. Even if the process does not fully meet the organization's needs, these past processes still provide a benchmark for end-to-end performance (either analytically by looking at data or by observing the processes in execution) and most likely can even provide insight into the cases where the process breaks down. You will be able to see where work accumulates in the process, which allows you to identify bottlenecks.

- We will cover **organizational change management (OCM)** in greater detail in a subsequent section, but it is worth noting here that being able to clearly map the current state to the target state for process participants is a very useful organizational change tool. This also unlocks the opportunity to index information in terms that people are already familiar with, which is good if you are producing content, such as quick reference cards, that people will refer to as they complete specific activities.

- Looking at current state architectures can also allow you to identify the system interfaces that can be preserved with their current interface definitions. This can significantly decrease your dependence on outside parties to execute unplanned efforts that would impact your project timelines if they were delayed.

- In the process of investigating the current state, you can often find useful tricks or methods in execution that could work as well in the future state as they do in the present. This might involve digitization of forms that are handled in Excel, or even physical forms, and can provide hints about which key data points to collect or which approvals to consider if you don't automatically treat them as the full specification of your target state system.

As you can see, there are many reasons to understand the current state, and learning about the current state to realize these benefits will provide some value to your project. Keeping in mind the limited resources of most implementation projects, the question then becomes how to balance priorities and enable a forward-looking view with an understanding of the past. The following strategies help you manage these tradeoffs and enable you to get more benefit per unit of time invested, which ultimately enables you to drive the maximum ROI given the constraints of your implementation. These strategies come down to allocating a small but significant fraction of resources to tasks that contribute to both understanding the past and shaping the future. By using these approaches, you can more efficiently determine how or where you should invest to get the most value and thus maximize your project ROI.

Strategy 1 – Process shadowing

When an implementation team starts working on a new ServiceNow project, some of the most common activities for the first week or two are to develop a project plan, standardize artifact templates, and otherwise set up the governance structure. While this is both useful and necessary work, there is another set of activities to which it is worth allocating a fraction of the time of any of your resources who are not actively spending every hour of the workday on the critical path activities. Process shadowing consists of becoming very well acquainted with the execution of the current processes by sitting with (or virtually shadowing) the participants of the process to get a deeper understanding of how they work today. Some examples of teams to consider shadowing depending on your process scope are the following:

- IT service desk teams
- Deskside support teams
- Customer support teams
- Field service technicians
- Data center support technicians
- Operators of related business processes and systems

This type of information gathering is particularly valuable as an addition to discussions and process reviews with the managers of these teams because many organizations show a disconnect between the documented process and the way work is completed.

Tips for conducting process shadowing

The following are a few tips for conducting process shadowing:

- Work with team leaders to identify the right individuals to shadow – ideally, you are looking for experienced and open-minded staff who are motivated to inform those working on systems they will ultimately use.

- Get two or three points of view if possible – the differences between how individuals get the work done can be as useful to know as the similarities.

- Explain in advance what you're doing and why; it can be disconcerting to have someone looking over your shoulder for an extended period, so having this communication done in advance, and preferably with the individual's direct manager, is advised.

- Observe and take notes of things you don't understand. Initially, you'll want to observe and not disrupt the flow of work. Resist the urge to re-engineer the process on the spot. This is your turn to learn from those who have spent hundreds, if not thousands, of hours executing the processes that you will need to update.

- Reserve questions for dedicated or natural breaks in the workday. Ideally no more than 4 hours should pass between opportunities to ask questions or clarify something you may have missed. This is long enough that most people will need to refer to notes but not so long that all context is lost.

- Note which auxiliary tools are being referenced during the work. For example, are mobile workers frequently using third-party navigation applications? Are service desk users referring to a documentation folder on the SharePoint site? Are Excel spreadsheets being used to look up assets or network resources?

- After the sessions, you should compile your notes into a summary of the current state process to preserve and organize the information for future review, as well as for comparison with the target state processes.

The time investment of process shadowing

With these tips in mind, you'll need to decide how much time to dedicate to process shadowing for your implementation. A good rule of thumb is to have 20% of your team spend up to one week conducting process shadowing. On a five-person team with a weekly total capacity of around 200 hours, this would translate to 40 hours of process shadowing.

As a typical implementation spans many weeks, the total fraction of project time dedicated to shadowing would often be ~1% of the total time. Naturally, if you are approaching the end of this period and you are still unearthing relevant insights, you may choose to extend this process. This is an opportunity to learn and improve the insight into the processes, not simply a checklist item to complete.

Another important consideration is who should conduct process shadowing. The best individuals to engage in process shadowing will be those responsible for designing the target state processes. On some projects, these are business analysts, functional leads, architects, or even senior developers or implementation specialists. Notably absent from this list are the project managers or project coordinators, as these individuals are likely fully utilized during the early project phases when process shadowing occurs.

The expected benefits of process shadowing

As always, activities should be conducted with a clear understanding of the objectives. In the case of process shadowing, you should be targeting the following benefits:

- Confirming that the full scope of the process is understood and included in the project's documented scope or charter

- Understanding the undocumented resources relied upon in the current process to ensure they are incorporated or replaced in the target state

- Identifying stakeholders for detailed process reviews

- Understanding how much behavioral change will be required from the users of the relevant business processes

- Identifying sticking points in the current process to ensure that the new process addresses those points

- Building relationships with the end users of the systems and processes that you're implementing and understanding their environment, mindset, and workload

Applying a limited amount of time early on in the project to process shadowing gives your team a rich resource to draw upon to become well acquainted with the current state without significantly increasing the total cost of the project. Process shadowing is an effective and low-cost hedge against the value trap of ignoring the current state.

Strategy 2 – Data analysis

Another useful and high ROI strategy allowing the current state to inform but not define your target state is to extract and analyze data from the current ticketing tools. This allows you to prioritize effectively and is complementary to the process shadowing strategy, as it will often outline outliers and patterns that would take weeks of observation to detect.

Targeting data analysis

While ServiceNow best practices often recommend not migrating ticket data in bulk, this data can still be a source of great insight into the current processes. Ideally, this information was used in the shaping of your project's value drivers but even if this is the first time that you are looking at it, there is still value to extract. Some examples of datasets that can provide useful insight include the following:

- Incident, case, and request ticket data from legacy systems

- Asset or CMDB data

- SLA data

- Knowledge Base article metadata (usage data if available)

- System user, groups, locations, and other master data

- Other datasets relevant to your project's value objectives

Often, some elements of data analysis are carried out on these datasets during preparation for data migration; moving this activity up in your project schedule and completing (or at least starting) it in advance of process design allows you to extract more value from roughly the same total effort investment.

Conducting exploratory analysis

Data analysis will start in an exploratory way and it is often useful to access the legacy system or process in a read-only form to see each data point in context rather than just as columns in a spreadsheet. Exploratory analysis should at a minimum get you familiar with the following:

- The time period covered by the data

- Key volumetrics of the data such as how many tickets, assets, and dollars hours

- Major categorizations such as regions, categories, teams, or request types

- Data quality, which fields are consistently populated, and where there are pervasive data quality issues

Integrating value and analysis

Once you have this information, you can combine it with your project's value statements to identify specific questions that should be addressed. While the precise process will vary, it is useful to explore a couple of examples to get a sense of how data analysis can be used.

If your project aims to optimize expenditure on end user assets, then confirming the number and value of the new assets purchased annually allows you to create a useful anchor point – it will also prompt you to consider the projected number of assets reaching the end of life in the next year (based on in-service dates and company policy, for example), plus the number reported missing or defective each year. If the rate of missing assets is particularly high, then improved asset tracking is likely to result in fewer missing or lost assets. In this case, improvements to processes such as employee offboarding or location audits should be considered.

If reducing the end-to-end cycle time in your software catalog is important, then your data analysis might focus on answering the question of whether all titles have a similar deployment time or whether some software is much faster than others (which prompts a discussion around why this is and how to address or leverage the differences).

> **Important note**
>
> Data analysis is intended to both answer some questions and raise new ones. You will not have the information to explain all the patterns and gaps, but being aware of them allows you to raise these during workshops or design sessions.

When analyzing ticket data, you will almost always want to group it by ticket type. This will both give you ticket volumes for each category (an essential metric) and allow a more detailed analysis of the peculiarities of each category. For example, analysis of all work orders for field technicians is generally less informative than separating them by customer installations, troubleshooting, or routine maintenance activities. You will want to consider what value your project is expecting to drive. If improving the efficiency of your field staff is important, then you may want to consider the time spent traveling, time on jobs, and repeat visits. However, if customer satisfaction is a more important metric for your project, then the **Net Promoter Score** (**NPS**) score from surveys may be more useful, allowing you to determine which work order types your organization performs well on and which frequently lead to customer dissatisfaction. The data alone will give you a part of the story and provide you with topics to discuss with the delivery teams and managers. You should use these follow-up conversations together with the data to determine where improvement opportunities lie.

The importance of intentional time allocation

As you can see from the tension between the last two value traps, there is tension between looking backward and moving forward with your implementation. The future is most critical but the journey to get there is informed by the past. As we have discussed, there are risks to letting the future state blindly emulate the past but also in ignoring the current processes.

The most important thing to take away from this is the importance of thinking critically about what your goals are to inform tough decisions on where to spend the most important resource in your project: time. It would be much easier to recommend a detailed analysis of the current state spanning many months, but few projects will have the resources to support this level of investment and even if they do, the time might better be spent on higher-value activities.

As a project team member or leader, you should always strive to be intentional in how time is spent on your project, asking whether the set of activities to be executed next supports the value proposition of your project to a greater degree than others that could be executed instead. This way of thinking carries forward to our next value trap, which fundamentally informs the allocation of time in support of value.

Chasing the long tail

The third value trap is based in part on a rule of thumb called the Pareto principle, more commonly known as the 80/20 rule. This rule states that, in most situations, 80% of the effects can be attributed to 20% of the causes. While the exact numbers tend to vary, value indeed tends to be concentrated within a relatively small subset of the possible scope for your project. This is most clear in the high-volume process configurations in ServiceNow such as those found in the Service Request Catalog. The value trap of chasing the long tail is experienced when the need to enable all instances of a certain workflow or automation type, regardless of the diminishing returns as a team, is applied to processes that are time-consuming to implement but are used only a few times a year.

The appeal of aiming for 100% coverage

No team or project charter sets out with the goal of working on a seemingly endless list of low-value activities – yet, many projects end up in exactly this situation due to unrealistic expectations being set at kick-off. Projects that get into trouble with the long tail are those that aim for completeness without considering the distribution of value across their scope. Typically, this happens when the scope is set at a high level early in the project (for example, discover all assets, or enable all service requests) and project teams then take this direction literally and execute it without considering the incremental value being delivered.

It seems intuitively clear that doing all of something is better than doing only 20%, 50%, or even 80%. Additionally, launching an incomplete catalog of service requests can lead to confusion among users looking for a missing item and this confusion can reflect negatively on your implementation. In this section, we'll look at the risk of this approach and provide practical guidance on how to avoid or address it. We will use the Service Request Catalog as a running example, as it is by far the most common case of this value trap, but the same principles apply to case types, discovery probes/patterns, and asset classes.

The distribution of value

Each process that you can implement in ServiceNow has an incremental value potential that it would add if configured in the ServiceNow platform. This value is determined by how many times the process is executed multiplied by the additional value that ServiceNow would provide if the process was effectively orchestrated by the platform.

> **Important note**
> Value potential is often achievable in stages – for example, a basic version of the process can provide 15 minutes of time savings while more extensive automation can save several hours.

The Pareto principle usually applies to these value potentials, meaning that a relatively small subset of the processes comprises a large fraction of the total addressable value potential.

Risks of completeness

Part of the challenge when addressing a scope focused on completeness is that every additional and well-implemented process provides some value or potential value to the organization. Given this premise, it is tempting to then suggest that implementing every process provides a net positive value, but there are two principal challenges with this conclusion:

- Given that all projects are executed in the context of limited resources, every hour invested in a particular process is an hour that is not being applied elsewhere in the project. This means that if work is being done on a marginally valuable process, then it is likely that something else that is potentially more valuable is being ignored elsewhere.

- When assessing the overall ROI, the potential value (the return) is one part of the equation – however, we should remember that implementation and support costs (the investment) should eventually be deducted from the potential value. Since there will always be some support or implementation costs, the actual value realized as a return on investment will always be lower than the theoretical value potential.

Taking these two challenges side by side, the risk of attempting to cover 100% of an arbitrary scope is that the result may distract from other essential or higher value activities and even the potential to invest more in the process than the possible future returns.

Unfortunately, it is far too common to see a team investing days into diligently holding workshops and producing documents to define a process that is executed only once or twice a year for limited value. If a process is discovered that truly provides an excellent ROI, then updating the business case or project scope to include it will also require additional time or a corresponding scope reduction elsewhere. The strategies that follow will provide useful approaches to reducing the risk that your project will allocate far too many resources in support of far too little value.

Strategy 1 – Top N selection

When defining the scope for an implementation where there is a potentially unlimited number of workflows to consider, it can be useful to set an arbitrary bound such as 10, 100, or even 1,000 processes and work within that bound to identify the most valuable ones to focus on. This approach requires an estimate of the number of valuable processes and the boundary number should not exceed this value. It is acceptable to set the bound lower than the likely number of valuable processes. It is often the case that completing the exercise to arrive at the top 10 or top 50 will provide much greater insight into the planning of a subsequent phase with a higher boundary number. In this way, the significant initial value realized can also help justify further investment.

There are many methods to select the top processes but they should be seeking to optimize value as defined for your project in all cases. Recall that the potential return on investment relies on three factors – volume, value, and investment. Theoretically, for a perfect ranking, you would need to define each process opportunity and assess these three factors and combine them to get a score that can be ranked. This takes a significant amount of time, so we will propose a modified approach that will be mostly correct but consumes significantly fewer resources.

The recommended approach for ranking

The following algorithm provides a systematic approach to ranking, which is one possible way to achieve a valuable list of processes for prioritization. It is not guaranteed to produce the best possible list but typically has very good results in most organizations:

- Set your boundary value for N (that is, 10 or 50).
- Of the three factors, the most reliably predicted is volume, as it can usually be established from a combination of historical data and business projections. First, determine the top 2N processes using the heuristic of the current state ticket volume. If possible, you may ask for quick adjustments by an informed manager (to account for known factors such as business changes or mergers that will have a major impact on any volumes).
- After ranking the top 2N processes by volume, you should now work with your technical team or architect to estimate a rough level of effort for each and with your business stakeholders to estimate the value of each. Re-rank the opportunities based on the ROI calculated using these factors.
- If you have at least N clearly favorable process investment opportunities, then you may use these as your initial N processes – if not, then you will need to repeat the process steps for the next set of processes by volume until you have a full list.

This procedure relies on N being smaller than the total number of positive ROI opportunities. If you find yourself running out of processes or assessing requests with very low value and volume, it may be an indication that N is too large, which would prompt a project leadership discussion to assess the value hypotheses of the project considering the new information.

Strategy 2 – Minimal implementation for long-tail items

For some projects, you'll find that establishing a complete catalog of services or requestable items is very important and that a low top N strategy would leave a gap that challenges the value of the catalog. In these cases, you can take advantage of ServiceNow's ability to develop a very low-cost implementation of a process using a reusable workflow to develop MVP workflows.

This strategy of a minimal catalog item implementation is intended to complement a Top N approach by considering the list of must-have items that do not fall within the top N high-value items. Instead of increasing their value, it strives to optimize the cost of implementation and maintenance to squeeze a positive ROI out of even a lower volume, lower value process.

> **Important note**
>
> This approach requires you to willingly adopt a standard workflow across these minimal implementation candidates, and a standard form layout with only a description, key header data, and a field to enter any additional details.

Applying this strategy populates new items into the catalog with a relatively low total effort investment and has the benefit of enabling metric tracking for these items in ServiceNow, facilitating more efficient analysis and optimization in future phases. When applying this strategy, it is critical to set expectations so that while some processes will be fully optimized, the ones that are implemented according to the MVP strategy will be more basic in form and function. Accepting this trade-off, this strategy allows you to get a large degree of coverage without overwhelming your team with the effort it takes to conduct detailed process analysis on each workflow.

Not managing change

A ServiceNow implementation almost always involves a significant change to the ways that people complete their daily work. This change typically results in a brief period of reduced productivity as workers acclimatize themselves to the new processes. Unfortunately, this period of reduced effectiveness can result in frustration and provide a negative first impression of the solution your team has worked so hard to implement. In addition, if gaps are not closed quickly, then the productivity hit can persist and permanently offset the value being realized from your implementation.

The first cut – OCM

When a ServiceNow project is in the initial planning stages, seasoned architects and project managers will typically highlight the need for OCM efforts to help facilitate the transition from the current to the future state. Unfortunately, when the initial budgetary estimates exceed the leadership's willingness to invest, one of the first areas targeted for cuts is the OCM effort.

This decision process is an exercise in prioritization and alignment with value. In principle, the decision to cut a lower-value part of the program would be appropriate – however, repeated experience shows that cuts to OCM are far more costly than most projects anticipate.

Risks of reducing the OCM effort

Cuts to OCM efforts come in different forms – the most common is to reduce the seniority of, and time dedicated by, the team members responsible for change management activities. In essence, this means reducing the cost of the OCM effort without fully removing it from the project. Another common approach is to restrict the time during which the OCM resources will engage in the project. These approaches lead to the following risks:

- Less experienced OCM resources take longer to become acquainted with the value proposition, scope, and implementation plan, and require additional support from the rest of the project team to effectively develop and deliver OCM efforts.

- OCM resources that are brought into the project significantly after kickoff often lack the context of the discussions that have occurred around the plans to adjust how things are done in the current state and thus cannot effectively map the journey of transition. Bringing these resources up to speed at a critical phase of the project (as go-live approaches) puts additional strain on the remainder of the project team during a period when many project teams are already fully occupied.

The impact of these two risks is a reduction of value realized across the implementation and additional strain on the project team at key times during the implementation and go-live processes. Recall that the most common reason OCM efforts get cut in a project budget is the desire to reduce cost and effort in an area where the effects will be less pronounced than if the project scope was trimmed overall. However, due to OCM's role in accelerating and securing value from the implementation, the effects tend to be more severe than expected. Fully and clearly articulating the role of change management as an accelerator to the full realization of the planned value is critical, particularly in combination with ensuring OCM efforts are rightsized from the outset.

Optimizing value from OCM

OCM truly acts as a multiplier for value from other areas of your implementation. That means that while it can have a large impact on the most valuable areas of your project, it is unlikely to generate value in areas where little impact is being made by the planned scope. This reality suggests focusing the OCM efforts on the areas where your project's impact is largest to ensure those are effectively covered and supported, and reducing the OCM efforts on lower-value processes. This approach requires you to consider OCM as a useful tool for value realization, not simply a checklist item.

Tying the OCM scope to specific value objectives also helps the budget holders visualize how OCM will support these value objectives and reduces the likelihood that OCM will be seen as a separate item that can be added and removed independently of the overall business case.

Responding to insufficient management of change

At times, it will unfortunately become necessary to recognize where OCM efforts have not been sufficient to prepare the organization for the coming change, representing a risk to value realization at go-live.

A red flag is a clear indicator that some kind of risk or issue is present in your project. Some examples of red flags that should be looked for as you progress in your implementation include the following:

- System users being surprised and confused during training sessions and acceptance testing

- Difficulty in producing target state operating instructions to cover the full scope of current working processes

- Poor awareness of the planned release of ServiceNow or poor understanding of how the release will impact daily work

Remember that OCM exists to ease the transition to new ways of working and to support the realization of value from those changes. If inadequate change management is evident in low-value processes but not in all high-value process areas, then that may simply be reflective of the tough prioritization decisions that were made. However, if your core value propositions seem threatened, then you will need to take immediate steps or risk significant portions of the planned value of your implementation being challenged.

Applying high-impact OCM activities

While the field of OCM can point to numerous benefits of the formal OCM process, the reality of ServiceNow implementation often requires balancing the need to do things right with the need to get things done. The activities discussed in this section will provide you with a toolbox of the activities that have shown themselves to be most valuable in ServiceNow implementations and that aim to reduce the post-go-live efficiency slump and user experience. The goal of these activities is to facilitate a smooth and sustained increase in value from the implementation and they have been curated accordingly.

Target state work instructions

Target state work instructions covering the current state process scope are perhaps the single most useful deliverable to facilitate OCM outcomes. Target state work instructions are produced on many projects already. With a little extra effort, you can ensure mapping and coverage between the current state of the processes that people use today to the correct procedures for their future work.

These instructions should be detailed and draw on the full range of common process scenarios (including those observed during process shadowing). The process of producing these instructions can be incredibly valuable because they cannot be created without detailed attention to the specific sequences of activities that users will complete in the system. While these work instructions do not need to cover every possible edge case, they certainly should be comprehensive enough to cover most of the cases that a worker will encounter in their daily interactions with the processes you are enabling on ServiceNow.

Transition support service

During the go-live and for a few weeks afterward, the establishment of highly available, knowledgeable, and friendly points of contact for users can provide a much higher degree of comfort for the teams working to get things done. The purpose of the channel is to allow workers or team leaders to get rapid answers to their questions so that system issues or gaps in knowledge do not impede their ability to efficiently complete their work. By providing real-time channels such as a walk-up help center, phone hotline, or Slack channel, you can immediately address questions and concerns and will become aware of issues minutes after they occur, rather than days or hours. This allows your team to react quickly and get an accurate pulse of how the transition is progressing and where the challenges are.

In an ideal transition where testing, training, and design have all been executed flawlessly, you will expect relatively little interaction with this supporting team, but it is still useful to deploy the channels, as the resources assigned to monitoring them can still complete other tasks during quieter hours. When a go-live is not as smooth as previously hoped, the extra capacity acts as a buffer, allowing your project to absorb some of the impacts on the organization and reducing the impact on operational teams.

The science experiment

ServiceNow technology is both flexible and powerful – this combination can lead to innovative solutions but also complex configurations whose cost to maintain exceeds their value. The value trap that we call the "science experiment" occurs when overly complex, advanced, or technically sophisticated architectures are layered onto ServiceNow, leading to substantially higher implementation and maintenance costs.

Science experiments are common in integrations, Predictive Intelligence, Virtual Agent, access controls, and even workflow business logic. The difficulty of recognizing a science experiment comes from the fact that they are often proposed by your most capable developers and the proposals do solve important design objectives.

Projects extending ServiceNow

There are many cases where basic ServiceNow capabilities must be extended for the efficient and effective realization of value. In some cases, the platform capabilities will get the job done but the configuration can feel overly burdensome, and a more abstract and efficient configuration layer is proposed. In other cases, an innovative new module of ServiceNow has been licensed and a mandate to implement is given even before specific measurable value objectives are defined.

To balance business needs with complexity, each ServiceNow project needs to determine to what extent it falls into each of the following categories: an implementation project, a software engineering project, or a basic research project. The nature of the project should be chosen depending on the value characteristics of the project and the degree of uncertainty in its technical execution:

- **Implementation projects**: These are by far the most common ServiceNow projects, aiming to deploy the system for maximum value and to remain close to one that is OOTB wherever possible. Implementation projects have higher success rates, a lower total cost of ownership, and execute on well-known design and development principles. Following the principles in this book, along with the ServiceNow technical best practices, provides a high probability of success in these projects.

- **Software engineering projects**: These ServiceNow projects are created when ServiceNow is chosen as a foundational platform with the expectation that significant custom development will be required to realize the expected outcomes. These projects may include the development of entirely new portal experiences, analytical capability, or deep integration into systems for which standard integrations are not available. Software engineering projects are more likely to experience delays and cost overruns than implementation projects, as the nature of the work is more variable and the full scope of the expected issues is not known in advance.

- **Basic research projects**: A basic research project doesn't aim to deliver something to production but rather aims to assess the feasibility of a certain idea or to evaluate alternatives for a specific problem. Very few ServiceNow projects intentionally operate as basic research projects, largely because the upfront commitments of ServiceNow licensing make projects without clear value outcomes and lower chances of success very risky. A basic research project is most often used in the opening phase of a ServiceNow implementation to assess the feasibility of implementing specific capabilities. A basic research project may conclude by proving the capability can be effectively implemented or even by proving that some constraint prevents the approach from being successful. Basic research projects do not need to deploy anything to production to be successful.

The following table shows the different types of projects and their associated complexity and success probability:

Project Type	Technical Complexity	Probability of Success
Implementation	Low	Highest
Software engineering	Medium	Medium
Basic research	High	Lowest

Table 2.1 – Project types

All three project types are potentially useful and valuable in their own ways, but it is essential to be clear on what type of project you are working on and to plan accordingly. The science experiment value trap occurs when elements of a higher uncertainty project are incorporated into a lower uncertainty project without accounting for the decreased certainty of outcomes. This disconnect signals a potential for a misalignment of the expected value with the value that will be delivered in your project – this misalignment is a source of potential risk that should be managed.

Risks of the science experiment

A science experiment can present several risks to your project:

- The experiment can consume a disproportionate amount of the senior technical resources' time on the project, preventing you from completing other objectives or responding to unforeseen challenges.

- Experiments are uncertain and as such, the effort or outcomes are variable. This uncertainty is of particular concern if a fraction of the project's committed value depends on work that is not certain to succeed or to result in a working solution within the project's planned timeline.

- Experiments can lead to highly complex configurations, even if completed within a scoped application. Future developers may be pressed to understand and maintain the configuration.

The risks of letting a science experiment run on your implementation or engineering project arise from the misalignment of the resulting complexity with your project's goals and risk tolerance. For some projects, it is possible that conducting some exploratory research and development efforts to support a significant and otherwise unachievable value target is the right course of action – however, this should be a deliberate decision taken with a full understanding of the costs and risks.

Recognizing a science experiment

Recognizing an unplanned science experiment early on is necessary to avoid the unexpected expenditure of significant effort that may not contribute to project outcomes. Some signs of a science experiment underway are the presence of the following indicators in an implementation project:

- Highly capable developers are working hard but without output in the form of value-contributing configurations.

- OOTB ServiceNow features are being passed up in favor of custom scripting. For example, the development of a scripted integration framework rather than the use of Integration Hub or the native data transformation function.

- Timelines for the work on one module significantly exceed expectations or are highly uncertain, particularly when the module is linked to more complex platform capabilities such as Predictive Intelligence or UI Builder.

When you suspect that a science experiment is being run on your project, then a discussion between the project manager and the architect or another senior technical leader about the value objectives, efforts to date, and value to date is required to determine the appropriate next steps.

> **Important note**
>
> A less experienced technical team may struggle with even more standard implementation activities. Ideally, each project will have at least one seasoned ServiceNow technical leader but if your team is all-new, the previous criterion may not apply. In these cases, you may need to rely on the advice of ServiceNow or a trusted implementation partner to help you assess the situation.

Handling a science experiment

If a science experiment is unexpectedly consuming the time of your project's resources, then it is necessary to align the objectives of the project with its execution. To achieve this, a combination of project management and technical leadership will be required. The value objectives leading to technical complexity should be assessed in the context of the current best information about the complexity required to achieve the value. If the business case to complete the work is strong, then commissioning a small software engineering or basic research initiative within your project should be worthwhile. This initiative should be managed with a greater eye to the risk and with the understanding that the timelines and success probability will have a different profile than typical implementation work. Again, this alternative complexity profile should be fully justified by the value that can be realized if the initiative is successful.

Summary

This chapter has covered five types of value traps that are common within ServiceNow implementations. We've covered the reasons why these traps are common and the strategies that can be used to understand and mitigate their effects. With the tools you've learned in this chapter, you should be able to strike a balance between designing for the future and understanding the business in its current form.

You will also be able to focus on a realizable and valuable scope that shows value to your project's sponsors to ensure that the trivial does not get in the way of delivering the critical. By developing a greater understanding of what change management is and why it is important, we have explored ways to justify the change effort and magnify its value. Finally, we have reviewed the risk of not taking a conscious approach to matching the technical complexity to the types of value being targeted.

Throughout this chapter, the recurring theme has been the alignment of effort to value during the scoping and execution of the project. In the next chapter, we will cover detailed considerations for managing and capturing value in a ServiceNow deployment.

3

Capturing Value from your Implementation

Now that you know what traps to avoid when it comes to value realization, it's time to highlight how and where ServiceNow can bring value to the organization.

We will look at how and where ServiceNow can help your organization capture value across the following themes:

- Lowering the process cycle time
- Optimizing asset utilization
- Automating repetitive tasks
- Improving customer experience

You will see as we elaborate on the various platform strategies used to achieve these outcomes that there are often various degrees of value realization that can be achieved depending on the scope and scale of the transformation. For some areas, the growth in value realization can be directly proportional to the investment, whereas others follow a more discrete pattern: investments provide little business returns until hitting a critical threshold upon which significant value is unlocked.

For each theme, we will summarize the value realization opportunity first, followed by what the common issues affecting value realization are in that particular area, and then detail how ServiceNow can help an organization capture the value. For each value realization strategy or tactic using ServiceNow, we will also provide some insights into how to plan for achieving the outcomes.

Lowering the process cycle time

Processes take a given input and produce an output of value to the organization. A lower process cycle time is about reducing the time it takes to complete the process from start to finish, with the primary goal of being able to run through the process more quickly and therefore produce value sooner.

Long process cycle times or unpredictable process cycle times can impact customer experience, as customer expectations are inconsistently met or missed. They also carry with them potentially significant financial costs where contractual obligations specify delivery according to a timeline with financial penalties attached to breached agreements. One area where long or unpredictable cycle times really matter is in the supply chain or procurement and asset management space where a long cycle time may mean very real increases in the financial risks, such as in the impacts on working capital or supplier payments.

Causes of long cycle times

Long cycle times do not simply just happen. Instead, there are many common patterns repeated across industries and markets, most of which arise due to the way organizations grow and operate. Learning what these common patterns are serves as a first step in determining how to solve these problems. We will list several such common patterns here that can help you identify how long cycle times may come about.

Confusion in how and where to engage a process

A frequent cause of long cycle times involves a lack of clarity on how a process can and should be engaged. This issue is common in organizations that are just starting to or have yet to build out their service management capabilities.

A sign of this problem occurring is when many services provided by IT or otherwise are engaged via the phone, email, or ad-hoc chat channels instead of through structured service requests. Typically, when there is a lack of clarity on how services should be engaged, customers will tend to prefer high-touch channels to be able to best articulate their issues and have a human help route them to the right place.

Having humans play the role of processing service requests and routing the work to the right place is only half of the battle (or less). If the processes and services being provided within (and outside of) the organization are ill-defined, even a person may not be able to appropriately determine where to send a particular request and what information to provide the receiving party. Another common pattern arising from this inefficiency can be seen in the sending of multiple internal emails and communications as part of a single request by a single requestor. Often, you will see emails going back and forth between the service requestor and the service desk, as well as the service desk and underlying teams who are trying to obtain the right information from the service desk and the customer.

If processes and the methods to engage them are ill-defined, the consistency of service request fulfillment times can be significantly impacted. This lack of consistency then results in several additional issues, which compound together to dramatically impact the organization. First, it is much more difficult to prioritize service requests effectively. This inability to prioritize will then result in more context switching in process and service operations, as tasks at risk of being overdue constantly appear and force the team to put all their effort into resolving them. This constant fire-fighting mindset can not only fail but even when successful, may also result in the highly inefficient delivery of service (either much earlier than necessary or perhaps just in time but with far more resources committed than should have been).

How to measure process cycle times and see your issues

It's easy to tell when process cycle times are long or inconsistent – it may be much harder to quantitatively measure and see where the issues are occurring and prioritize the opportunities according to the return on investment.

Recently, multiple products have arrived on the market that can help the organization better understand and identify improvement opportunities. These 'process mining' systems plug into major service management, order management, or supply chain management platforms, automatically ingest and interpret the data flowing through these systems, help create a visualization of the health and flow of the process, and identify any bottlenecks.

If you suspect long cycle times are impacting your processes, utilizing a process mining solution in a targeted way to identify bottlenecks may allow you to take a more targeted approach to resolve the issue, improving your overall return on investment.

Delays in response to requests for service

Response delays can impact process cycle time by straightforwardly delaying the start of process engagement. Several factors can cause response delays. First, if there are simply too many incoming requests or inquiries compared to the throughput of the process, new inquiries and requests will begin building up in the backlog, drawing out the period between the time of submission to when the process is formally engaged.

Another cause for long response delays occurs when the intake channel for new requests is not formalized (as in the previously mentioned confusion over how to engage a process). In this case, the service fulfillers may not be monitoring the intake queues, especially for infrequently engaged channels – this can then result in long delays before a process is formally engaged.

A lack of upfront prioritization capabilities may exacerbate any existing response delay issues. When queues are backed up, it pays to work on the most urgent requests, inquiries, or issues first, but if there is no clear system to prioritize work, then a first-come-first-serve (or last-come-first-serve) model may arise very naturally. In the best case, the randomness of service requests causes the average response time to even out, but in the worst case, this can cause significant delays, as frequent but low-value inquiries and requests block the rapid allocation of resources for servicing high-value but less frequent requests.

Inability to find the right expertise to fulfil the request

Even when there is a structured way to engage in a process, the fulfillment of that process may still involve the identification of exactly where to send it after the initial attempts to address the service request have failed.

This situation occurs more commonly with processes that are initially engaged without perfect information, such as incident or problem processes. In these process types, it may not initially be clear what kind of expertise is needed to solve the issue, or indeed, even what the reason for the issue may be.

With these types of service requests, the initial response time may be rapid, but the resolution time may be exceptionally long, as the service fulfillment teams struggle to send the request to the teams where the cause or root cause can be identified, the issue resolved, or the request fulfilled.

Lost updates and poor transitions between service fulfillment parties

Processes often involve many different parties and some processes may even involve third parties external to the organization, such as contractors, vendors, and external service providers. Process cycle times may be significantly impacted if these third parties are delayed in delivering their results.

In addition to delays or failures to deliver by these third parties, cycle times can increase because there is no immediate feedback on when these external parties have completed their work. There are two primary reasons why this may occur: first, there may be no clearly defined way for the vendor to communicate the completion of their work back to the originating service, or the method to communicate back may be deficient in some way. A common example of this is when a particular third-party fulfiller is engaged via an external support system or channel. The fulfiller, upon completion of their service, updates a service request ticket in their system, at which point the onus is on the originating process fulfillment team to check that system to determine when the fulfillment has been completed so that they can take the appropriate next steps. The delay between when the process step has been completed and when an agent checks that system can significantly impact the process cycle time.

Second, the completion of the external party or process may not have clear traceability back to the originating process. This can be a much more significant source of process delays compared to the first issue. A common example of this is in hardware asset management or software asset management processes. A customer engaging in the process of getting a new laptop may encounter a situation where the fulfillment process identifies a lack of stock. The team fulfilling the IT asset request may then need to engage procurement to purchase more stock, which then needs to be shipped by the vendor to the shipping dock and delivered to the appropriate stock room before being sent to the customer. By the time the laptop has arrived at the shipping dock, the team receiving the laptop may not have a straightforward way of identifying that this laptop received was associated with the initial order, and therefore the update that they make to their inventory system to indicate this laptop has been received may be invisible to the initial service request or lost. In such a case, it is common to see the request be completely forgotten about until the customer, frustrated, calls in about the status of their order, which is then finally fulfilled as the agent checks the stock room and finds the laptop in stock.

How your ServiceNow transformation can reduce process cycle times

ServiceNow can reduce cycle times of processes through the following platform capabilities, often working in conjunction with each other.

Hosting structured service requests forms that may be driven by fulfillment workflows

The Request Management capability in ServiceNow, when combined with Service Catalog and Workflow, allows you to provide a centralized location where users may request services, reducing any confusion about how services and processes may be engaged.

Formalizing the request intake using electronic forms can reduce the number of low-quality requests that are received by the fulfillment teams. By making requests for service (and the engagement of processes) highly differentiated, the intake forms can contain error checking and input checking so that the user is encouraged and incentivized to provide the right information and the fulfillers can provide service immediately.

Finally, structured and differentiated service requests can then have workflows that send the requests to the right teams and individuals to service the requests at the right time. Workflows can also trigger automation and integration and send out appropriate notifications to agents if they do not consistently monitor the ServiceNow platform. With consistent processes, workflows can drastically reduce response delays and lost updates by facilitating the transfer of a task from one team to another.

The value that can be achieved by simply identifying frequently engaged services and putting them onto a central portal where users can engage these services using clear and simple forms should not be under-estimated: for organizations that are used to service engagement through email and phone calls, it can singlehandedly improve the customer experience and reduce the cost of service.

Utilizing master data such as the CMDB to automate or reduce the time spent on often time-consuming manual procedures

ServiceNow workflows and automation can be made significantly more powerful and effective with well-managed and purposefully designed master data. Master data, in this case, refers to data within the system that is consumed and shared across capabilities and processes – it is also often referred to as foundational data. The **Configuration Management Database** (**CMDB**) is a classic example of master data in that it is shared by multiple IT service management processes such as incident management and change management. Capabilities and processes utilizing master data can realize value beyond what can be achieved in isolation, as the master data itself 'links' the processes and allows for activities and events occurring as part of one process to signal other related processes. For example, utilizing shared application service data combined with technical service data events detected by event monitoring systems can automatically trigger the creation of incidents and the assignment of incidents to the technical service team closest to the probable cause. Taking a step further, the team investigating the incident may then look at change records associated to **Configuration Items (CIs)** with a relationship to the source of the error to help them narrow down the possible change that may have caused the problem.

Connecting multiple processes to prevent lost updates and improve on-time fulfillment rate

Shared master data and shared processes operating on a single platform can improve or eliminate the problem of lost updates during process transitions. An example is hardware asset management, enabling the teams at a loading dock to check in assets that have been purchased previously in the procurement or request process and notify the waiting party that their shipment has arrived.

ServiceNow bridges these gaps by tracking some aspects of each process and connecting them by looking at specific state transitions or data changes in one process that is shared in another. For example, in the case of hardware asset management, ServiceNow tracks hardware assets across multiple processes. When an asset is initially ordered, a record is created that associates the asset with the purchase order. Subsequently, using data from the vendors, the asset is associated with a serial number, which can then be scanned in at the loading dock to identify that the asset has now arrived. Upon arrival, ServiceNow can then notify the customer requesting the asset that they will soon be able to receive the item and have the asset moved to the right stockroom for distribution.

How to prepare for an implementation with a focus on reducing the cycle time

When attempting to use the ServiceNow platform to reduce cycle times, make sure you have prepared in the following ways as part of the implementation to improve your chance of success.

Clearly understanding the major process bottlenecks and the value of removing bottlenecks

Before starting the actual platform implementation, it is helpful to identify through process mining or other techniques where the actual process bottlenecks are. Once bottlenecks are identified, it is good to prioritize the bottlenecks by order of the value returned when the bottlenecks are reduced or removed to focus the implementation's objectives and scope better.

Identifying exactly what kind of value you are looking to achieve with process optimization will help the team determine how best to solve the problems they will inevitably encounter during the implementation. Without an identified objective, it will be easy for the team to get lost and go beyond the scope of what is required or miss the ultimate objective as they get bogged down in the details of the specific issue they are facing.

Identifying and aligning the stakeholders and parties that will need to be involved to complete the change

If a bottleneck occurs due to lost updates or poor management of the master data then process optimization may require the involvement of multiple parties and stakeholders, often more than what you might initially expect due to the interconnected nature of typical enterprise processes. For example, optimizing hardware asset management processes often requires the service desk, the supply chain, and the IT, procurement, and HR departments to be involved and make changes. This does not even include external vendors who may also be involved in the end-to-end process.

If one or more of these groups are not aligned or are unwilling to implement changes in their respective processes or procedures, the results of the transformation may be significantly impacted. Therefore, an important first step in the implementation is identifying the types of change that may be required of these stakeholders and obtaining alignment with these stakeholders on the right timeline and resource commitments required to enable them to make these changes.

Now that we have covered how you can leverage ServiceNow to reduce process cycle times and identify where these cycle times may be, we will move on to look at another use case well-aligned with the capabilities of the platform – the optimization of asset utilization to reduce overhead and costs.

Optimizing asset utilization

Asset management has traditionally been an area where organizations invest substantial manual effort due to the lack of technology and the complexity of coordinating the various functions of the organization involved in the management of the asset life cycle. The problem is complex enough that for some organizations, certain asset types, including crucial ones such as end user computing devices, are completely neglected.

Low asset utilization and lost assets can cause tangible and serious financial impacts on an organization. The optimization of asset utilization includes increasing asset reuse, increasing the overall utilization of an asset's value, identifying lost assets, and improving the purchasing or leasing strategies of assets so that they can be procured as cheaply as possible to maximize the return on investment for these assets.

A further important benefit of asset management in the IT space is in security. Understanding the disposition of technology assets – in particular, end user devices – can reduce the chances of security compromises by enabling the organization to quickly recover assets that are no longer needed by employees (such as in the case of offboarding) or be alerted about the possible loss of assets.

What do you consider an asset?

Any organization will have many types of assets, from office chairs and application servers to buildings, and many of these assets are managed by different teams, so there may be disparate requirements on how they should be managed.

When discussing the optimization of asset utilization in the ServiceNow context, we primarily mean IT assets (servers, laptops, desktops, software licenses, and mobile devices), as ServiceNow has out-of-box solutions created for managing these types of assets.

Within the vast universe of IT assets out there, only a proportion of them will be 'managed' by the organization. These are assets that the IT department or organization has determined to be important enough (typically due to the total cost of ownership of the asset) that the disposition, procurement, and utilization of these assets must be monitored. Not all IT assets should be managed, as the return on investment when tracking certain assets is low – examples of these types of assets include most consumables (e.g., mousepads).

While the focus in the following content will be on IT assets, many of the concepts apply to other types of assets within the organization. ServiceNow can also be used (and has successfully been used) to support the management of assets outside of IT, but these may require additional configuration.

Causes of poor asset utilization

Poor asset utilization can be caused by loss, over-allocation, or over-procurement, amongst other reasons. In this section, we will list many common causes to help you identify possible opportunities within the organization. It is important to know that while you may find poor asset utilization using these criteria, you must also qualify the opportunity. Not all assets are created equal and likewise, not all poor asset utilization scenarios are worth solving. With that said, it is also true that when it comes to IT hardware and software, there is often at least one or two asset classes whose poor utilization can incur significant negative costs to the organization.

Lack of a clear view of the organization's asset inventory

To utilize assets effectively, organizations should have a clear view of what assets they have and where those assets are located. While this sounds simple in theory, in practice, this visibility is obscured within an organization by process, governance and technology gaps.

First, there may not even be a clear understanding of what kinds of assets are to be managed and what kinds are not to be managed within IT or the organization. The lack of formal standards in this area will often result in a lack of standardized processes to keep track of assets, leading to immediate losses of managed assets, as some assets slip through the cracks.

Software Asset Management (**SAM**) is a notable example of this issue. Without a formal SAM team or function within the organization, software assets tend to be purchased by many different departments for various purposes and tracked in different ways (or not tracked at all). This lack of standardization or a central purchasing process can make it impossible to determine the full extent of the organization's software asset inventory. Software assets are particularly difficult to manage, as they are often entirely intangible, and unlike hardware, assets cannot be physically counted.

Without a clear view of the asset inventory, it is impossible for an organization to efficiently decide when and how to purchase new assets – it also significantly impedes opportunities for asset reuse, which then leads to increased operational costs for the organization.

Inability to determine the actual disposition of assets

Even if the asset inventory is well managed and understood, organizations may still have trouble maximizing the utility of assets because once an asset is known to be owned by the organization, the exact disposition of that asset is lost.

For example, laptops may be purchased, placed into a master inventory list, handed out, and then lost until an inventory count finds the laptop missing from a stockroom or facility.

Some of the difficulties in managing the disposition of assets include the fact that not every step of the asset life cycle is managed by the same teams in the same systems, and in other cases, it can be extremely time-consuming or resource-intensive to manage the state of these assets, even if a centralized group of individuals decides that they would like to track them.

Inability to understand the true utilization of an asset

Ideally, an organization wants every asset purchased to be highly utilized, with the assumption that high utilization of the asset contributes directly or indirectly to business value. It can be difficult, however, to measure what the utilization of an asset is and even what constitutes good utilization versus poor utilization.

For many high-value asset processes such as IT hardware asset management and IT software asset management, measuring useful utilization metrics is typically an activity that must be done using automated systems and tools. However, even when these tools are readily available, integrating them and federating them into a comprehensive view may prove to be a challenge.

How your ServiceNow transformation can improve asset utilization

ServiceNow has two primary modules that are directly created to optimize asset utilization: Software Asset Management and Hardware Asset Management. At a very high level, each module provides conceptually similar functionality: the ability to track the inventory of assets, the ability to track the disposition of assets, the ability to track the utilization of assets and the ability to integrate with other systems that may bring in asset data across these areas.

When evaluated in detail, software and hardware asset management are very different products as the high-level concepts of asset management translates into extremely different practical problems that each product is purpose built to solve.

ServiceNow provides a suite of capabilities that when set up and working together can be extremely powerful in achieving asset optimization goals. Initially, there are a set of three broad capabilities that should first be established:

Tracking IT hardware assets across the life cycle

ServiceNow's Hardware Asset Management function allows the organization to track the inventory and disposition of IT hardware assets across their entire life cycle, from request and purchase to retirement. When processes are aligned to utilize this system, the organization can have a single view of its IT assets, where they are located, who they are assigned to, and where they came from.

ServiceNow's IT Hardware Asset Management capability is also well integrated into the CMDB and can improve the CMDB by providing initial IT hardware CI data before the CI is installed into the IT environment. In this way, IT asset management provides an entry point for some CI types, giving processes such as Change Management and Release Management an early heads-up that these resources are available and may be used as part of implementations.

When service requests are designed to take advantage of IT hardware management, requests for IT hardware may directly utilize real asset inventory data in their fulfillment processes. Should procurement be an activity in these processes, procurement can utilize IT asset data to optimize purchasing decisions (e.g. by buying in bulk whenever possible) that can reduce the total cost of ownership of IT assets, thereby improving utilization.

Tracking IT software assets across the life cycle

Similar to IT hardware asset management, ServiceNow also provides capabilities to track the life cycle of software assets available to the organization. Unlike IT hardware asset management, however, identifying just what constitutes an IT software asset and 'counting' how many are available is a far more significant task for software assets. Because software is intangible, it can be difficult to even define what is the 'asset' component of the software, let alone being able to track its state.

ServiceNow's software asset management capabilities provide organizations with reference models for software assets for major enterprise software manufacturers. These models allow the organization to manage software assets correctly, sometimes using named user licenses, and sometimes using several available installs. ServiceNow also provides capabilities to detect these assets, as they are installed across IT infrastructure, provided ServiceNow's discovery capabilities are running in the environment or the data is being integrated from external systems such as Microsoft SCCM.

ServiceNow's software asset module also provides ways to both automatically and manually track the state of software assets once they are in the system. Again, unlike IT hardware asset management, the intangible nature of software means that there are many different ways to track the disposition of software assets. For named user licenses, each named user consumes a purchased asset, while for utilization-driven licenses, the number of entitlements (or assets) may be depleted as they are installed. To complicate matters even further, the depletion of licenses may be based on other properties such as the number of CPU cores on the server a software may be installed on. Without software asset management, manually managing all the various types of licenses and entitlements that are available can become almost impossible.

Tracking hardware and software utilization

Determining the utilization of assets first requires that 'utilization' be appropriately defined. Subsequently, depending on how utilization will be measured, processes and technology can be put into place to monitor and improve that utilization.

For end user hardware, such as laptops and desktops, metrics such as login times are common and easy to obtain, as popular configuration management systems such as SCCM can be used to automatically report these metrics. ServiceNow can integrate with the system as a native platform capability and collect utilization metrics against the assets.

Software asset utilization is more difficult to measure, as usage information may not be directly available or if it is available, must be measured using proprietary approaches. For software covered by its publisher packs, ServiceNow has pre-built capability to determine usage. This can greatly simplify or reduce the amount of additional technology to be integrated with the platform to track utilization.

In any case, the actual tracking of utilization itself is typically done with technology outside of ServiceNow, but the platform provides capabilities to integrate with these data sources.

Once these three initial foundational capabilities are established, additional investment into the following three areas can substantially increase asset utilization and return on investment on the platform.

Reducing software asset compliance costs and reclaiming software assets

Poor software license management not only costs the organization in the form of penalties but over-purchasing can also be a source of substantial expense for the organization.

Once the capability is in place to identify and track the software asset inventory and measure the consumption of entitlements and the utilization of these entitlements, ServiceNow can then produce reports and alerts for when the organization's software compliance position is at risk or has been breached.

The same synthesis of collected asset metrics for compliance purposes can also be used to help the organization reclaim unused or underutilized software assets. This can be particularly useful in the case of highly expensive, consumption-based licenses, where reclaiming an underutilized entitlement can save the organization tens of thousands of dollars per year for each reclaimed entitlement.

Optimizing asset procurement strategy

When purchasing assets, do you purchase a single asset as a one-off – or perhaps purchase in bulk to try to save money via economies of scale? When purchasing in bulk, how much of a particular asset should be purchased so that there is just enough inventory on hand for operational needs and not too many in storage?

When requests for assets, purchasing, receiving, distribution and utilization can all be monitored on a single platform, the ability to make asset purchasing decisions more strategically and in a more optimized way becomes possible.

With this information, ServiceNow can be used to centralize asset sourcing, pool requests for assets, and enable asset management teams to purchase assets in bulk by utilizing asset consumption trends to decide exactly how many assets to purchase.

Identifying and reclaiming hardware assets

For end user hardware assets in particular (laptops, for example), tracking the life cycle and utilization of assets from the time of distribution and installation provides organizations with the opportunity to reclaim potentially lost or under-utilized assets.

There is great potential for the loss of assets during the employee offboarding or transfer processes, either when employees are provided with new equipment as they move between departments or geographies or when employees leave the organization. By keeping track of what each employee has been provided with during onboarding, the offboarding process can utilize this information to reclaim the assets of a departing or transferring employee. Additionally, tracking the utilization of hardware assets through metrics such as login time allows for asset management teams to trigger automated attestations to validate that the asset is still being used and available.

Even if reclamations do not successfully occur, identifying the number of hardware assets that are lost annually can be a valuable metric to drive future continuous improvement opportunities.

How to prepare for an implementation with a focus on optimizing asset utilization

While ServiceNow can bring significant business value through optimizing asset utilization, careful preparation is needed to improve the success rate of implementation.

Before engaging in any asset management implementations, it is recommended that you are clear on the following points.

There should be a clearly defined asset management process owner

Asset management requires at least some level of centralized control and governance. There should be clear accountability in terms of who will be managing the asset data in the ServiceNow platform and the overall value realization of hardware asset management.

Sometimes, assets are managed in a distributed manner in an organization due to organizational silos (for example, perhaps IT manages most end user IT devices but the business manages permanent workstations in certain office locations). The asset management process owner can, but does not need to be, from any of those groups and can be part of a separate organization altogether. The asset management process owner does not necessarily need to be a direct manager of all individuals managing assets, but they should be someone at a someone empowered with sufficient authority to be able to set and enforce process standards for teams managing assets and ultimately report asset management metrics to the organization.

The asset management process owner should work with the executive team of the organization to clearly establish the scope of the asset process they are managing. This may include establishing factors about the metrics such as the types of assets that will be placed under management, what must be tracked to determine the success of the process (e.g., asset spend or asset utilization), and the escalation path for the process owner if they identify organizational changes that are outside their direct authority.

Is it also possible that an organization requires multiple asset process owners to manage all of the assets that it deems 'important'. For example, it may be much more effective for the asset process owner managing end user computing devices for an organization to be different than the process owner managing corporate capital assets, such as buildings or large equipment. In the IT-specific context, you may still choose to have multiple distinct asset processes by splitting across organizational lines such as regions or lines of businesses. From a CIO or COO perspective, if these distinct asset owners exist, their overall effectiveness is best managed through having common reporting metrics as opposed to focusing on transforming the distinct operational processes into a single unified whole. For example, while the EMEA region may largely lease their laptops and the APAC region may largely buy them, the process owners responsible for IT hardware asset management in both organizations must collect metrics on the dollars spent per laptop and the business customer satisfaction with the IT hardware provided in the exact same way. These common metrics will align the distinct process owners in fulfilling the same important business objectives while providing them with the flexibility to optimize according to their local environment.

In summary, when allowing for multiple asset process owners in an organization, it is important to make sure that they each have clearly defined domains that have minimal overlap in terms of accountabilities and responsibilities, and this generally means splitting responsibilities along clearly distinct asset types (hardware versus software and facilities equipment versus IT equipment) or clear organizational lines.

Defining the scope of asset management based on the area of value to be realized

Just because it's called *asset management* doesn't mean that the scope is everything that could be considered an asset within the business. Assets can be tables, chairs, computers, servers, or transistors on a motherboard. Not all assets are worth the effort of tracking and optimizing. Additionally, what may be worth tracking tomorrow is not necessarily the lowest hanging fruit today, so careful consideration of the immediate scope of which asset utilization you wish to utilize compared to what you wish to manage in the long run is important to define beforehand.

Without a clearly defined scope of exactly what kinds of assets will be managed and why we want to manage them (that is, which of the value realization opportunities mentioned previously we are looking to take), it can be very easy to under- or over-commit resources during the implementation, as the scope either creeps beyond what is optimal or the stakeholders, time, and resources allocated are insufficient to truly create a positive return on investment.

When defining the scope of the implementation, make sure the following is clear:

- Exactly what assets will be managed by the system and supporting processes.

- How the implementation will deliver value (for example, which of the points in the *How your ServiceNow transformation can improve asset utilization* section are you going for? Which of the problems in the *Causes of poor asset utilization* section are you looking to resolve?)

- What supporting processes and teams will be subject to (or must execute) business change as part of the transformation and which ones will be off-limits (for example, procurement, supply chain, or the IT service desk).

The third point is particularly important, as implementation without the involvement of the right stakeholders might produce a half-baked solution that doesn't realize the full potential of the platform in optimizing asset utilization.

Aligning and ensuring that all teams and stakeholders to be involved in the transformation are prepared

As with any other transformation initiative, the alignment of stakeholders is critical to the success of an asset optimization or asset management implementation on ServiceNow.

For asset management, consider the readiness and alignment of the following stakeholders and teams (depending on the implementation):

- **IT service desk**: Usually involved in asset management, as they may be the main team in the distribution of hardware assets and managing the installation of software assets.

- **HR**: Specifically teams involved in the onboarding, offboarding, and transfer functions that may currently provide end user hardware and software assets.

- **Procurement/supply chain**: Depending on the organization, asset purchases may go through centralized procurement or supply chain teams, and optimizing spending will require procurement to buy in to adjust their processes to leverage the new tools and data.

- **Facilities/receiving/mailroom**: Whoever may be involved in the initial receipt of assets at the loading dock and who (depending on the target process) may be involved in entering assets into the system once it has arrived.

- **Vendors**: Vendors may provide advanced asset data before the asset arrives in the environment, allowing traceability early in the asset life cycle. Vendors may also be responsible for asset tags or serial numbers that can be used to uniquely identify assets. For certain software assets, vendors may have tools and technologies that monitor asset utilization and asset usage, and these tools and technologies must be integrated with ServiceNow to complete the view of the software asset in the environment.

- **Independent business units with significant off-the-shelf asset purchases**: In many organizations, business units or business leaders may have access to individual budgets and may make substantial asset purchases. Depending on the scale of this purchasing, asset management may need to involve key business units and leaders so that these purchases are either consolidated into centralized processes or tracked so that enterprise-level optimizations can be achieved.

When embarking on an asset management transformation, you may encounter situations where some groups just do not wish to be a part of the change. Even in the most hierarchical and integrated organizations, business realities will still exist that may create resistance between a business unit, group, or team in following some broader integrated process to support an enterprise goal.

For example, it's possible that a particular team has extremely high local levels of maturity and there would be an unacceptable impact on a team or its customers if they simply adopted a less mature enterprise process (e.g. the engineering team has a hardware procurement and provisioning process that's highly automated with a small number of preferred vendors that they set up in isolation so that they can deliver their releases at a rapid pace, far beyond the needs of the rest of the organization).

While it may be possible to accommodate these differences, to match maturity levels, or myriad other 'ideal' options, there will be situations where the constraints of resources, time, and competing priorities may truly cause the integration to be impossible.

In such cases, the transformation initiative should look to reduce its scope so that the impact of non-involvement by the particular business unit, team, or group is minimized. This means not only restricting the scope of the target state process from the team 'out-of-scope' but also ensuring that there will still be sufficient business value delivered if the out-of-scope team does not change their process, procedures, or technology at all to accommodate the implementation of the target state transformation. Transformation plans should be halted and replanned if the return on investment is no longer justifiable without the involvement of the team. This should not be seen as a negative and instead, seen as an opportunity to identify why discrepancies in terms of priorities, initiatives, or level maturity exist and pivot the transformation to these areas first before proceeding with a broader asset utilization optimization transformation. The leadership of the **organizational change management** (**OCM**) team is crucial in determining whether these options should be considered, as they can provide a clear and people-centric view of the level and legitimacy of resistance and help the project's steering committee to decide whether the costs of removing the resistance are worth the business value ultimately obtained.

Optimizing asset utilization involves the addition or enhancement of new and existing business processes to take advantage of the platform's automation capabilities. In the next section, we will look at leveraging the automation capabilities of the platform to eliminate the boring, repetitive, and high-volume activities that erode productivity.

Automating toil

There are numerous repetitive, heavily repeated tasks in the operations of a business – wouldn't it be nice if some or all of these tasks could be automated somehow so that a human could go do something more creative or more valuable with their time?

Sources of toil

Automating toil is an easily understandable concept and a major source of business value that can be delivered by the ServiceNow platform. While there are countless areas of work that can qualify as *toil*, given the areas of focus of ServiceNow, the following are several major relevant areas to focus on.

Creating, assigning, and managing tasks

As boring as it sounds, one of the most basic, time-consuming acts of toil is simply the management of tasks and to-dos, including the creation of them, the assignment of them, and keeping track of a list of them so that nothing is missed.

The management of a large queue of tasks can be tedious and error-prone if performed quickly, and when these tasks are arriving in an unstructured format such as email, time must usually be taken to convert the tasks into structured formats in systems so that they can be reported on and tracked.

To get a sense of how much time and effort is spent on this activity, examine the number of unique instances in a particular process flow where unstructured data is interpreted by a person and converted to structured data and where a record of a task (an email or a ticket) needs to be interpreted by a human and sent to another human or back to the customer. For each action that exists, assume a reasonable amount of *action time* (such as 30 seconds) to get a sense of the amount of time consumed by this simple activity alone.

Copying data or actions across systems

Organizations with complex workflows or many distinct teams and processes that interact with each other may also use multiple distinct technologies. While it is not wrong to choose technology that is fit for purpose, issues arise when some teams act as a bridging point between multiple technologies used by different teams as part of the same process.

For example, a front-line customer support team may take calls from customers and need to dispatch field agents to the customer site for additional triage. The field services team and the customer support platform may be different platforms, requiring an agent to copy information already captured in one system to the other to inform the field team that they must be dispatched. This discontinuity between systems may also add additional manual work if status checks are required – for example, if the field agent did not make it to the site, the customer may call for a status update, requiring the customer service agent to look for the customer on in the customer support platform first before then checking the field services platform to see why the visit was not accomplished.

This type of swivel-seating can be extremely time-consuming. Each time data is copied, not only is time spent to copy the data (upward of 1 minute or higher) but each transition can also cause errors and information loss that impact the quality of service delivery.

Operational reporting and distributing operational reports

Operational teams can spend hours or tens of hours every week compiling operational reports. The less structured the incoming data is, the more time-consuming and less accurate these reports become.

Operational reporting should be differentiated from analytical reporting. The former is meant to provide a more granular (and ideally real-time) view of the immediate situation. Operational reporting is useful to help operational teams allocate resources, manage staffing levels, and identify operational gaps and issues. Analytical reporting is often done for longer-term, predictive use cases such as determining long-term organizational strategies and computing metrics that may only become apparent when data from many sources is aggregated and then transformed to be useful for a particular algorithm or heuristic.

Operational reporting is generally a greater cause of toil (and certainly a greater opportunity to reduce toil) than analytical reporting because it is repetitive, typically comprised of numerous simple but manual steps, and occurs on a fairly frequent basis. Operational reporting's most time-consuming aspects include extracting any required data into a tabular format, cleaning up any data in the reporting that may be invalid or incorrect, interpreting the data line by line, applying metadata that may be needed to facilitate operational reporting metrics, calculating the operational metrics, and generating reports for the metrics.

In addition to the work of actually compiling the report, another time-consuming process concerning operational reporting is the distribution of the reports to stakeholders, involving identifying the stakeholders, crafting the communications, embedding the report files into the communications, and sending it.

Regression testing

A ServiceNow platform that is thriving as an enterprise enabler of value will inevitably experience a high volume of changes as new configurations are released regularly to deliver additional value to the organization. Inevitably, the pace of these changes runs into the competing issue of maintaining the integrity of the platform and reducing disruptions caused by the unintended impacts of change.

For smaller implementations or inactive platforms, regression testing existing functionality by hand may be feasible. For large enterprise implementations where releases occur frequently and ServiceNow is used as a platform for numerous business-critical teams, automating the testing of existing functionality not only reduces toil significantly but also significantly improves time to value. ServiceNow provides automation capabilities for regression testing using its **Automated Test Framework (ATF)**, which can significantly reduce this overhead.

Where ServiceNow can automate toil

ServiceNow provides several key capabilities that can be used to automate the various sources of toil within service delivery.

Workflow engines, email actions, chatbots, and structured service requests for automating the toil of creating, assigning, and managing tasks

ServiceNow has a collection of capabilities that when used in conjunction with good process design and underlying master data, can significantly automate the toil of creating, assigning, and managing tasks.

The actual principles of this automation are simple: common customer requests are analyzed so that the standard set of information needed from the customer to deliver service is structured into purpose-built intake fields on an electronic form. Automated error checking can then be implemented to reduce errors from customer entries. The structured form will improve the quality of the requests and improve the chances that the request can be serviced correctly the very first time. The agent's toil of having to engage the customer to truly understand what they are asking for is also reduced by this improved clarity.

Once the request has been submitted, the structured input data can then be used as a component of automation. ServiceNow has a workflow engine that can direct the task to the appropriate teams and individuals for service based on both the intake data and the business logic working on that data. For example, an HR service request for a recommendation letter may go to two different departments depending on whether the requester is a contractor or a full-time employee. With the appropriate HR data integrated into the platform, ServiceNow can decide this on behalf of agents and route the request to the appropriate team automatically.

Sometimes, even though a structured intake exists for a particular kind of service, customers may intentionally or unintentionally avoid engaging with that structured intake due to ignorance, convenience, or preference. Whatever the case, ServiceNow provides capabilities such as Virtual Agent that can help take these unstructured inquiries, prompt the user with the appropriate questions, and then turn the initially unstructured engagement into a structured request. In this way, ServiceNow can help increase the number of customer engagements that become structured requests that can then be targets of automation.

Integrations for supporting e-bonding and sharing data across systems to reduce swivel-seating and copying data between systems

ServiceNow makes integration with other systems relatively easy to build and maintain. Components such as its Integration Hub and its native integration and API capabilities can be used in a fit-for-purpose way to automate the copying of data across systems at critical transition points.

Several use cases are particularly low-effort but high-value – for example, when records need to be created in remote systems (such as when customer service agents have to dispatch field agents using a different system) and only the remote agent's text and status updates are needed to help the ServiceNow agent obtain the appropriate visibility to best serve the customer. In these cases, having a 'shell' record on the ServiceNow platform reflecting status and work-note information from the target system using integration can provide the ServiceNow agent with the required visibility while requiring very little technical complexity.

The remote ticket can even be created from ServiceNow via an integration where ServiceNow passes the appropriate information required by using the initially created ticket, reducing the need for agents to take the same action twice in two different systems.

Some caution should be applied to attempting to create 'true' e-bonding designs when automating task copying across systems. This refers to e-bonding solutions that attempt to merge business processes across two systems using technical integrations, which can be very complex to accomplish and should be avoided when other alternatives exist.

In most use cases where ticket information needs to be passed between systems, the appropriate design is to create a proxy record in ServiceNow representing the basic information about the state of the remote record and link that proxy record to the initial task record (for example, an incident created in ServiceNow is passed to a remote team on a different platform on an incident task record, which serves as the 'proxy').

With this design, the proxy record's state and other information can be independent of the parent record, creating a great amount of flexibility on the ServiceNow developer's side in terms of how to change the parent ticket's attributes given changes occurring to the proxy ticket (if at all).

A design that should be avoided unless absolutely required is a true 'e-bonding' system where instead of creating a proxy ticket to designate the remote ticket, the initially created record is, upon a triggering action, effectively synchronized to a record in a remote system and from that point, effectively exists as a virtual mirrored object across two systems.

The risks of this design are that unless the target system can fully mirror the source system, and unless the target system's teams execute identical processes and procedures, the synchronization process will need to be complex to translate the various idiosyncrasies between the remote and source records. For example, if the remote system has five incident states while the source system has four, the e-bonding will require logic to translate the odd state on the remote system into a valid state value on the source system, or the remote system will have to block the extra state when the ticket is in a synced state. If all that sounds confusing, it is because it is. E-bonding solutions should be avoided whenever possible in favor of the proxy record design and even if attempted, it is recommended to use technologies specifically created for this purpose to manage all the issues that could arise from such a design.

Before embarking on any e-bonding solution, your platform team should make sure it is familiar with the various out-of-box capabilities offered by ServiceNow to simplify e-bonding in specific cases. For example, Integration Hub offers spokes that provide out-of-the-box capabilities for querying, updating, and creating tickets in supported external systems, which can reduce the need for bespoke integration development. The Service Bridge capability of the Telecommunications Service Management application provides a framework for facilitating e-bonding between ServiceNow instances.

Real-time operational reports and dashboards eliminate the time needed to create operational reports by hand

ServiceNow has out-of-the-box, competent operational reporting capabilities that can be used to generate real-time operational reports based on the data flowing through the system.

ServiceNow also provides the capability to present these reports (and related reports) on real-time dashboards, which can be made accessible to specific individuals and groups.

This operational reporting capability, when appropriately leveraged and planned for during the implementation of the platform, can eliminate or drastically reduce the need for manually created operational reports. If some may not want to access these reports and dashboards in the system, ServiceNow can schedule these reports to be sent to specific groups and individuals regularly by the more traditional email channel.

The automated testing framework can be used to increase agility and reduce the manual effort of regression testing

ServiceNow's ATF can be used to automate testing platform changes. The ATF is a built-in testing tool that can exercise ServiceNow platform changes, including interfaces built on the service portal.

The ATF is highly complementary to test-driven development approaches where tests are written to fail and then platform changes are implemented to pass these tests.

Once automated tests have been set up to test a particular functionality, platform deployments (including minor releases) should require test suite runs to verify that changes in a release have not impacted the expected behavior of the existing configuration.

While automated tests can significantly reduce toil in the medium and long term, they do add overhead to the configuration process. The time it takes to write good tests that are robust and fail for the right reasons should not be underestimated, and teams should expect an increase of at least 10 to 20% in the overall development effort (some say it could be 50% or greater) when automated tests are written as a rule for all configurations. Despite the costs, the rewards of automated testing are significant, both in terms of long-term savings in toil and reducing risk to the business caused by bad changes.

How to prepare for an implementation with a focus on automating toil

Besides the universally relevant truisms 'understand your stakeholders' and 'understand your use cases', there are some preparations to make when it comes to planning and preparing for an implementation targeted towards automating toil.

Measuring and ranking your sources of toil

One of the most important preparations when looking to automate toil is to measure and rank your sources of toil. This means collecting operational data, either manually or automatically, to determine where the greatest source of toil is occurring so that your automation initiatives can target the right automation opportunities.

While there is likely to be a broad array of places where repetitive and menial tasks happen, the greatest opportunities are typically found in areas where there are high volumes and even small improvements or savings can lead to dramatic savings in time and frustration.

Another easy way to identify potential areas of toil is to simply ask the people that do the work – they will likely provide a host of insights into where toil happens and opportunities to automate exist. Whenever stakeholders are questioned or interviewed, the information should always be used only as a starting point – some level of measurement should be done to quantify the actual impact on the business and enable prioritization.

Standardizing processes and procedures

Automation is not magic. Machines are great at automating tasks with a fixed number of objective and quantifiable decision points and not so good at making decisions that rely on subjective interpretations of ambiguous data. This is true even today when advances in artificial intelligence have made machines much better at making decisions based on larger input data sets with less discrete differentiation between those inputs.

The first step in enabling automation in areas where repetitive actions or manual tasks are currently being performed is to attempt to standardize or simplify those actions and tasks into clearly defined logical workflows and decision-making points so that the rules can be captured and then executed by the platform instead. To do so, you should examine the processes of opportunity and design changes into the process to better facilitate automation. This usually means identifying the distinct decision points that need to be made, standardizing the input data required to drive those decisions, and standardizing the actions taken once those decisions are made.

This concept of designing or changing business processes in support of automation is not only powerful for preparing for ServiceNow implementations but is also a generally applicable concept that can significantly differentiate a business in its ability to return investment on technologies.

Clearly understanding how value will be measured for returns

While any initiative should be measured for its returns on investment and value realization, the automation of toil is a particularly strong use case for having quantifiable measurements of value because it's usually clearly aligned with directly measurable metrics.

Establishing ahead of time how value will be measured and return on investment calculated can help focus the implementation on dealing with toil in the right places and the right ways, avoiding overcommitting or under-committing resources or over- and under-pivoting on requirements.

One important pitfall to avoid when it comes to measuring the return on investment on this type of initiative is to think that the only way to actually realize value from automation projects is when there is a reduction in headcount. This perspective is shortsighted and neglects other benefits that could be just as or even more valuable than operating costs but are simply not being regularly measured – improvements to the quality of service, improvements to the quality of life of your staff (leading to reduced turnover), and enabling your teams to focus on more creative work, leading to new business value opportunities being generated that no one had time for in the past.

Up until now, we have extensively discussed opportunities for the platform to deliver value to the organization by optimizing and enhancing its capabilities to deliver service efficiently. In the next section, we discuss how these capabilities, in combination with others, can help create exceptional customer experiences.

Improving customer experience

Regardless of how well an organization has optimized its processes and streamlined its services, these improvements are irrelevant if they do not tangibly affect the ability of the organization to deliver a great customer experience. In this section, we will highlight the levers of customer experience that can be pulled to improve it and how ServiceNow plays a role.

Where customer experience can improve

Whether it's an IT organization at a small organization or a globe-spanning product company servicing millions of customers, the improvement of customer experience is an oft-pursued goal when it comes to ServiceNow transformations.

The issues that organizations are looking to resolve typically follow a few key themes.

Providing a non-fragmented service experience and a single place of service

Service portals and digital storefronts are all concepts related to solving a common customer service problem – how to provide a user-friendly and personalized place where customers can engage the services offered by an organization (or parts of the organization).

The concept itself seems simple enough – if a single highly personalized and user-friendly portal exists for the customer, it will improve customer experience and reduce customer frustration when they need help or support for one or more of these services. It also allows the teams supporting that customer or delivering that service to focus their resources on improving that single portal or storefront, increasing efficiency and generating greater returns on investment.

In practice, customer service remains highly fragmented. Within the enterprise, there are more often than not many 'storefronts': IT service portals, the company intranet, and so on, and it can often be highly confusing for employees to navigate the maze of options available to them. If no clear guidance is provided, business customers will resort to what they know best: reaching out by email or phone to get the support they need, not only increasing their frustration but also the operational costs of the business.

While the consumer or customer-facing space tends to be in better shape, many problems still exist. While there is typically only a single digital engagement channel initially, the problem is complicated when acquisitions and the launch of new services fragment that storefront.

A single view of the customer across all service channels

For product or service organizations with consumers as customers, the concept of a *single pane of glass* or a *single view of the customer* describes the ability of agents and systems to get visibility into a customer's history, preferences, and past interactions regardless of which channel the engagement has occurred within. A typical example of this is for a customer service agent to be able to see that a customer has had a recent interaction with a sales agent inquiring about new services or that several returns have been processed in the last 6 months.

The single view of the customer ideally enables agents and systems to present services, options, and support in a more personalized fashion, specifically in the world of consumer and services industry organizations, which could lead to greater monetization opportunities.

For the customer, a single view can also lead to more consistent service experiences and engagement across various channels (imagine going to a walk-up or store and being provided tailored services or experience because of your frequent online purchases and preferences).

In the context of enterprise services meant for the internal organization, a better internal view of the customer can also allow more targeted enabling services, better prioritization and stratification of support levels, and better analytics to help enterprise services deliver better outcomes for their customers. An example of having a single pane of glass in the internal organization includes tying HR, facilities, and IT services together to enable employees to have a seamless onboarding and offboarding experience.

Linking customer-facing services and products to internal services to predict service impact and plan service improvements better

Customer-facing organizations often struggle to connect their internal business services to their customer-facing services despite the deep relationships that often exist between the two. For example, telecommunications companies rely on both technicians and support personnel facing the customer to deliver service, as well as internal technical teams who maintain applications, servers, databases, and core network infrastructure that delivers telecommunication services to the customer.

In these deeply connected organizations, the ability to see the interdependencies of these disparate parts of the organization can improve customer experience by better managing the risk of change impact. It can also deliver better service by letting customers know when downtimes may occur and when service may be restored. Again, using a telecom example, in the case of outages caused by issues in the core network or internal applications responsible for the delivery of service, customers often have little visibility into even the most basic information, often because it is difficult for even the telecom to understand which customers may be impacted by the outage they are seeing. By creating more visibility into these dependencies between internal and external services, it becomes possible to provide customers with more proactive and regular feedback, improve their experience, and reduce the number of calls to and engagements with the customer support organization.

The visibility of the interconnection of services can also help product-driven organizations to coordinate marketing, sales, product design, and customer support better and reduce intra-company friction. Sales and marketing often love to create as many product and support options as possible to optimize monetization opportunities, whereas the support and product organization prefer more simplified options as it reduces the complexity of the support and maintenance. By connecting these domains through integrated processes and data, the organization has a better chance at finding the right balance that optimizes cost against revenue generation.

Where ServiceNow can help improve customer experience

Now that we have seen where several areas of customer experience can be improved, we will look at where the platform can add value by improving the experience across these areas with its capabilities.

Consolidating services into a single portal or related portals on a single platform

ServiceNow's Service Portal capability provides a way to reduce the number of integrations necessary to tie multiple distinct service delivery groups together, consolidating a fragmented customer experience into a more cohesive one.

The advantage of using ServiceNow as your platform to deliver a service portal or portals is primarily gained through the ability of ServiceNow to also be a work management and workflow management tool in the delivery of these services. Many dedicated portals or content management solutions will require more complex and difficult-to-maintain integrations with the underlying systems of action to provide users with visibility into the status of the services they have requested from the portal – ServiceNow can provide this level of visibility in a much more technically straightforward way, with significant reductions in the number of custom integrations required. This tight integration enables organizations to provide a deeply interconnected and cohesive experience with fewer resources upfront as well as reduced operational costs over time.

How to make objective decisions on designing a subjective customer experience using experiments

Taste and preferences vary by individual. Organizations pursuing a transformation to consolidate or create a brand-new service portal or service catalog may have questions such as *"How many portals should we have?"*, *"Should we split our services into their own distinct portal areas or have them all on a single page?"*, or *"What should our catalog look like and what categorizations should we apply?"*

Many of these questions are highly subjective in nature and can be prone to contentious debates that go nowhere, increasing the cost of the transformation and stalling time to value in the worst cases.

The secret to success is embracing the fact that there will *never* be universal agreement on something so subjective. Instead, the 'best' solution will often be the solution with the greatest buy-in despite the existence of reasonable counter-perspectives. With that said, a democratic voting process is not always the only option when deciding which design is best. User experience design tools such as A/B testing, card sorting (and reverse card sorting), and mouse and eye tracking heatmaps can be used to provide more objective data on how designs differ. In the case of methods such as card sorting and reverse card sorting, it is possible to decide with the appropriately sized testing groups which proposed catalog categorization structure is objectively better than another, even if the advantage is slight.

Integrating end-to-end service delivery processes and channels onto a single platform

Enabling a single view of the customer often begins with consolidating the work products of operational processes that engage or support the engagement of those customers into a single cohesive design. ServiceNow, by virtue of its focused capabilities for workflow management and task management, can serve as the primary platform of work for many of these channels, with its robust and easy-to-maintain integration capabilities bringing in further data and insights from other systems of engagement to provide a single, unified view of the customer.

The ability for ServiceNow to take advantage of federated data comes from using that data to drive workflows, improve operational reporting, and assign the right teams to help the customers at the right time.

ServiceNow also provides a *'next-best-action'* capability that allows rules to be created using the customer data to guide agents on how to provide the best support and service outcomes. This capability is particularly important as the amount of data known about the customer (and related to the customer) grows. A human will find it difficult, especially under time pressure, to be able to consume all the relevant data and decide on a best course of action, but having the platform provide these recommendations on behalf of the human agent allows them to take full advantage of the federation of customer data and provide a differentiated service experience.

Tying internal and external services together using the Common Service Data Model and CMDB

While the concept began in IT Service Management, the concept of the CMDB and ServiceNow's implementation of its CMDB and related capabilities provides the building blocks necessary for internal operations and customer-facing operations to understand their interdependencies better and create a clear picture of how they impact each other.

When following the common service data model design and with clear use cases and governance in place, ServiceNow can be leveraged to help organizations anticipate, visualize, and measure the true impact of internal service outages and the problems caused for the customer and the cost of maintaining and supporting the new products and services sold to the customer.

Being able to visualize and tangibly see how these processes and services relate and connect can often act as a starting point for more ambitious business transformations that will truly align the organization from end to end. A valuable goal would be to create services and products that are not only optimized by looking at sales but also by incorporating the total costs and requirements of support and operations through the lifetime of the new product.

How to prepare for an implementation to improve customer experience

When implementing a project with a significant customer experience improvement objective, one of the most significant steps to take is to identify a set of objective measures alongside more subject measures of value realization.

Capturing a baseline of the relevant metrics and then subsequently measuring improvements in these metrics when the transformation is underway or complete could in and of itself be a project or transformation. This is especially true in the case of customer experience when most metrics need to be collected against a large representative sample size to be relevant and prevent overfitting or focusing on something that feels significant but is not in the grand scheme of the overall customer experience.

Knowing where to focus customer experience improvements is another important activity that can help with implementation success. Key elements of this understanding when it comes to a customer experience focus include the following:

- Clearly defined types of profiles for customers that describe distinct or unique needs and behaviors
- Collections of user journeys that these customer profiles undertake, ranked from the most common journeys to the least common ones
- An idea or directional alignment on which journeys need improvement (ideally backed by measurable data points)

Summary

In this chapter, we looked at common value realization opportunities within organizations and how ServiceNow's capabilities can be used to capture these value opportunities. We also looked at how careful pre-planning can make transformations more successful when they occur. Common themes that enable success for all value opportunities include establishing clear governance, identifying metrics and measurables to create quantifiable goals and objectives, and having concrete ideas on what should be changed in the target state compared to the current state.

In the next chapter, we will look at a broader collection of these optimizing factors that you should be aware of so that your transformation has the greatest return on investment and chance of success.

4
Planning an Implementation Program for Success

No matter how much you know about what value you wish to capture and what objectives you would like to achieve, a program is only successful if you execute it properly.

This chapter takes a thorough look at the various dimensions to consider when planning your program implementation, the common failure scenarios (and how to avoid them), and the roles each member of the implementation team should play (or not play). You will come to see that a successful implementation is never about how spectacularly the project solved a single problem or decisively made a single decision. Instead, successful projects are constructed from a combination of clear vision, clear plans, and checks and balances. In this chapter, we will cover the following:

- Defining the scope of the project
- Planning the release and release activities
- Structuring the implementation team
- Governance
- Value management

We will see how these activities contribute to the ultimate success of your implementation. Within each dimension, we will highlight common techniques, patterns, and methods. We will also discuss how the project team, both internal and external to your organization, should contribute to the success of the activity, and common pitfalls to avoid.

Defining the scope of the project

Defining the project's scope is a critical part of the success of your project's implementation. Without a clearly defined scope, your ability to measure whether the value of the project has been achieved will be made more difficult. In addition, the project team and any governance and decision-making structures created for the project will be significantly less effective at making decisions and taking action. Scoping a project should not just be about defining what should be done. Well-crafted scope includes insights gained from careful planning that can guide the project team in understanding possible risks, better prioritizing areas of uncertainty, simplifying solutions, and avoiding over-stretching their resources on out-of-scope activities (i.e., scope creep).

Important components of the project scope

There are many other places where you can find guidance on how to define a project scope. We will therefore highlight the aspects that are specifically important to ServiceNow implementations to reduce the amount of overlap between perspectives.

For any project, the following components are critical to be defined within the project scope documentation.

The project vision

When this project is over, what will have changed? A project vision can provide a concise summary of what the project is meant to achieve. The act of creating the project vision is also a useful litmus test for seeing whether the key stakeholders and project sponsors are aligned at a high level.

Project visions should be concise and allow the project team to quickly determine, "Does what we do contribute to achieving the project vision?"

When creating the project vision, it can be extremely helpful to articulate the vision in terms of changes to the current states of particular stakeholders. This technique allows us to create vision statements that starkly highlight the change that the project is expected to deliver. Taking an approach of highlighting anticipated stakeholder changes can also help avoid the common pitfall of having a vision statement so general that it fails to provide any useful guidance in practical terms. For example, "increase the automation of IT support" is of less practical use than "IT customers requesting for services will require fewer steps than before to request for their services."

The project's guiding principles

Guiding principles are a set of objectively measurable statements that can help determine how to make the right decisions to achieve the project vision.

Project guiding principles should be written in such a way that if every project decision meets the requirements of the guiding principles, then the decision should almost be guaranteed not to negatively impact the realization of the value of the project. Guiding principles can often be even more critical to define than detailed functional requirements from a scope perspective, as they are much more effective at enabling teams to deal with uncertainty, a key factor in delivering a successful implementation.

It is important to use guiding principles that are clear in their application. For example, a guiding principle of "always think of the customer first" sounds great in principle but is practically less useful at helping guide decision-making than a guiding principle that may say "the project should never make decisions that increase the number of steps taken by a customer to obtain a service," or "the project should never make a decision that increases the average time to service of the customer."

This is not to say that more generic statements are to be entirely avoided, as they can sometimes be helpful to orient the project team toward the right dimensions to consider. In the previous example, the more generic guiding principle may still be valuable if it is to be used to help an operations-centric team make decisions that are more considerate of the impact on customer experience.

During project execution, additional guiding principles that target workstream level nuances (e.g., having an additional set of guiding principles for a customer portal branding and experience) may be created, but for the purposes of scoping, a single set of concise guiding principles to inform the entire project should serve as a sufficient starting point.

Example guiding principles include the following:

- Always reduce the number of unique process steps across teams rather than configuring technology to accommodate these differences.

- Never increase the time that a service takes or the number of actions taken by a customer before they can receive service.

- Reduce the required agent decisions and agent actions required to fulfill service requests for a customer.

- Don't put off making changes that impact the organization – deliver the needed organizational change along with the systems.

- All proposed changes must respect the project scope, timeline, and resource constraints.

Current state processes in the scope

Having a list of current state processes that will be in the scope of the transformation is a great starting point for identifying the stakeholders involved and perhaps finding stakeholders that you may not have been aware of in your initial ideation of the project. When identifying processes, it's helpful to inventory the organization's actual operational processes instead of using ServiceNow's module processes or the **Information Technology Infrastructure Library** (**ITIL**) processes. While these standard process lists can serve as a useful starting point, they may be too generic to capture the full scale and scope of the transformation that the organization needs to undergo to achieve value.

For example, in many organizations, *incident management* in the ITIL sense can be several operationally distinct variants with the same high-level objectives. If a dimension of value is to consolidate these distinct sub-processes (and disparate technologies that may be supporting these distinct processes), then capturing the specific number of distinct incident processes in operation can be an important step toward truly understanding the scale and scope of the transformation.

The preceding example of distinct variants of a standard process such as *incident management* may sound absurd but is not uncommon, especially if your goal is to use ServiceNow as the single incident platform across multiple parts of the organization.

Typical situations where you may encounter operationally distinct variants of the incident process include any organization with customer-facing teams versus internal teams (e.g., support teams for customer-facing technologies operating distinctly from internal IT support), or organizations with very distinct IT groups due to factors such as historical acquisitions or disparate geographical locations.

Identifying and placing appropriate guard rails around the scope of change in these cases can be critical to the success of the project, as attempting to merge these variants into a unified process (or accommodating these variants in the platform) can be time-consuming and will be made even more difficult if the appropriate stakeholders have not been initially engaged in or brought into the transformation.

The project's desired value and KPIs

As much as possible, it is important to decide upfront how the success and value of the project are to be measured. It takes preparation to do so, just as showing improvement in the target state requires that you first obtain a baseline of the KPIs. To further complicate the process, the way a KPI is measured in the current state may be different from the target state if a significant process change occurs as part of the implementation. It is therefore important when defining the success measures of the project that the scoping team do the following:

- Establish a set of KPIs to be improved as part of the project

- Define how the KPIs will be measured both in the current state and in the target state

- Take a baseline measurement of the KPI in the current state, possibly requiring previously uncaptured data to be captured

Establishing KPIs is a lot of work but is often worth the effort. It not only provides the project with an objective measure of success but also works as a way of focusing the transformation: requirements or proposed changes that do not contribute to the improvement of the target KPIs can be de-prioritized in favor of changes that do.

A functional requirements list

If possible, it is always good to capture and include the functional requirements for the target state in the scope to provide an upfront level of detail for the implementation that can help better identify potential risks of the project. Functional requirements can also be used as a tangible way of measuring the completeness of project implementation. Well-written functional requirements can easily be verified against the target state implementation to see whether each has been fulfilled.

A common pitfall of creating functional requirements as part of project scoping is taking an approach of collecting requirements by committee. Sending a functional requirement template to various stakeholders with little to no oversight of how the individual requirements fit together into a cohesive, working process can lead to a confusing and conflicting list of requirements that will be difficult to streamline into a working solution during implementation. It is therefore recommended that if requirements are to be built ahead of time, a team comprised of at least one seasoned platform architect and supporting functional designers is leveraged to help organize the functional requirements into high-level working design ideas.

Furthermore, tying functional requirements into the inventory of current state processes can bring even more clarity to the desired outcomes for the implementation team. Diagramming these workflows in a swimlane diagram or another visual format will help to align the processes and can avoid costly technical rework down the road.

A technical requirements and integrations list

As ServiceNow implementations take place within an existing platform with out-of-the-box capabilities, technical requirements at the time of scoping should focus on specific capabilities that require special accommodations not found directly in the out-of-the-box platform capabilities. Some examples of such considerations include the following:

- Performance requirements, large data volumes, or near-real-time response requirements for large data volumes

- Integration requirements – in particular, integrations that have to go through **Enterprise Service Bus (ESB)** systems or have to conform to enterprise integration standards

- Encryption requirements that specify exactly what kind of data is to be encrypted and the type of encryption required (at rest, in transit, or end-to-end)

- Accessibility requirements, such as accommodating visual impairments (while ServiceNow has some accessibility capabilities, out-of-the-box requirements in this area should be evaluated specifically)

The scoping of technical requirements for a ServiceNow implementation should take an informed approach to avoid over-specifying requirements that are provided for by the platform out-of-the-box or are beyond the capabilities of the platform. For example, ServiceNow transparently recommends that the system ought not to be used for fully real-time applications or applications with complex graphical rendering requirements (e.g., producing 3D images), and any should trigger an evaluation of alternative technology choices.

Adding details to the scope

When scoping the project, a balance should be struck between how much detail to include in the scope of work versus how much should be left open-ended or high-level. There are multiple trade-offs involved and there are no clear answers on exactly what is the right level to provide to the implementation team to deliver. There are, however, a set of guidelines and rules of thumb that, if followed, can provide consistently good results.

Having the internal project team flesh out the details of the project scope can be more effective than bringing in third-party expertise. While external experts can help facilitate the discovery of these details, they are often limited in the advice they can provide for areas that may be highly dependent on the specific goals, objectives, or organizational model of the business.

In other words, third-party consultants may be able to provide the techniques and tools for understanding the project but are less efficient at collecting the actual details themselves because of additional time required to form a complete understanding of organizational nuances. Therefore, an important function of collecting the following details is to help all the parties involved to be more effective by understanding the landscape of the scope of work and its known risks and uncertainties:

- **Detail risks and uncertainties**: If the solution to a particular objective is uncertain or unknown, it is better to include information on where the uncertainties are instead of hypothesizing about the possible solution. This is doubly important when the implementation will be carried out by a third party (e.g., a consulting company), as it allows that third party to provide better estimates taking into account the risks and allows project managers to better mitigate risks upfront and plan for contingencies. A good example of this approach is when the system will rely on a dataset of unknown quality. Effort may be required to cleanse the data and this work can be difficult to forecast. Documenting this uncertainty prevents it from being forgotten and ensures that the project continues to track it as a risk to successfully realize value.

- **Detail KPIs and measures of value**: A repeated theme in this book is the focus on value and measures of value. A project is far more likely to demonstrate value if there are clearly understood, quantifiable objectives that can be baselined before the project and measured again after the project. Whenever possible, a substantial portion of resources and time during the initial preparation of the project scope should be focused on what quantifiable measurements will be used to determine project success and to guide project priorities.

  When defining KPI measurements, it is not enough to simply list the names of the metrics and then move on. Instead, how the KPI will be measured should be clearly defined, and measurements before the start of the project should be taken to formulate a baseline that can then be improved upon. Determining what and how to measure could often at a glance seem intangible or difficult to do. It is a science on which entire books have been written, and it is highly suggested that project leadership begins with careful planning and prioritizing these measures, either internally or by consulting with external experts:

- **Detail the stakeholders, groups, and decision-makers**: The act of defining the stakeholders, decision-makers, and groups impacted by a project can help clarify the scale of the scope of work, particularly in large transformations. For example, an IT CMDB project that involves the data center or infrastructure maintenance groups will differ significantly in its size and the possible outcomes it achieves compared to one in which they are not involved. In the latter case, the transformational outcomes should be more limited to enabling the automatic detection of application-level faults and dependencies, as, without change or commitment from the infrastructure team, any desired hardware-level dependency management in the CMDB will likely not be operationally sustainable.

Mapping this list of stakeholders and decision-makers against the in-scope processes and sub-processes can provide the final layer of insight to guide the project team in designing and implementing the target state changes to meet the project objectives. As we will see in the subsequent sections of this chapter, understanding the decision-making landscape of the project is a critical tool in accelerating and improving the project's ultimate outcomes.

Planning the release and release activities

With the scope appropriately settled, it's time to think about the second aspect of successful implementation: determining an appropriate timeline for your project that accommodates the various releases and release activities.

The obvious reason why a timeline should be established for the implementation program is to attempt to gain certainty over when the value and returns on investment will be realized (or begin to be realized). A less obvious reason is to use the timeline as a way of establishing constraints that can help the project team and the organization focus more clearly on outcomes and goals.

Due to the inevitable existence of uncertainty and risk with any project, planning the timeline and release strategy can often be an art form. Some level of uncertainty and risk is inevitable, but it is possible to use a series of guidelines and guiding principles to estimate the timelines of a project and determine the appropriate release strategy.

In this section, we will present a list of considerations, strategies, and techniques to be considered as part of your release and release activity planning. It is important to keep in mind that determining how long and how to structure these release activities often comes down to an exercise in risk management. Allotting more time to ascertain that each activity has been completed to a high degree of detail and quality will reduce the risk to the implementation and value realization. Moreover, occasionally taking reasonable risks by reducing time, or parallelizing activities, more often than not, may enable the organization to realize the value more quickly. While the balance of risk versus reward is something that must be weighed against each project differently, the strategies and activities to consider are consistent and therefore summarizable in the following sections.

Releasing often (minimizing impacts)

In the spirit of agile methodologies, projects should release whenever value can be realized by the release. This is frequently summarized as *"release early and release often"* in product development. In the enterprise transformation and major enterprise platform implementation space, there are additional complexities to consider in the form of negative business impacts caused by the replacement of an older but more complete solution for a newer but more limited solution.

This impact risk can be particularly acute with the first release of a major enterprise platform meant to replace an existing solution – for example, ServiceNow replacing an existing **IT Service Management** platform, or ServiceNow replacing a customer service management ecosystem. In such cases, there are often significant legacy capabilities and processes with significant dependencies upon each other. A ServiceNow implementation targeting only a small subset of these capabilities may cause either an unintended impact on the *out-of-scope* processes or be forced to make undesirable concessions in target state design to maintain the operation of those processes.

Therefore, for enterprise deployments, effort should be spent on determining the point in time that brings both net new value to the organization and has a reduced impact on dependent but out-of-scope processes. This does not mean the release timeline should simply wait until all legacy capabilities have been replaced. Instead, it means that for very complex organizations and environments, more time and resources should be invested in creating short to medium-term interim states. This will help bridge any gaps a new deployment may initially cause to dependent but out-of-scope processes to enable earlier releases of value in specific areas of the project.

For example, an HR service management implementation may be able to quickly bring new automated HR service requests with substantial cost savings on a new HR service portal, but it may take significantly longer to replace all the components of the existing HR portal. In such a case, it may still be worthwhile to have an interim state where both portals exist, but strong communications and modifications on the legacy portal will be put in place to direct employees to the new portal and its automated services in the interim. This allows the release of the automated functionality earlier without impacting employees who still need access to the capabilities on the old portal that are not available in the new one yet.

Planning releases to maximize value and minimize impact extends not only to when releases can happen but also to where. For large global deployments, for example, it might make sense to release the target state regionally if the impact of change can be isolated. In HR service management projects specifically, regional release opportunities can often be found. This is because, due to regional legal requirements, even the most unified HR organization can operate independently at the regional level.

Considering external dependencies

Do not neglect dependencies on external groups and teams when planning the release. As part of the release plan, consider the availability and readiness of the following:

- **The external vendors involved in managing or delivering services**: The availability of vendors providing managed services for technologies or infrastructures that the project will need to change or provision, or vendors who are involved in a current state process that will be changing in the target state, should be considered. Dependent vendors should be contacted early on in the planning process and incorporated as key stakeholders in the project and project timelines. This dependency can have significant impacts on the release schedule, as vendors are often bound by contracts that may prevent them from changing quickly at the discretion of the project.

- **The teams involved in external systems that the platform will be integrating with**: For ServiceNow implementations, integrations with external systems are frequently encountered and utilized for automation, obtaining business data, and other business purposes. Integrations with these external systems will require coordination with the teams supporting them in activities such as design, development, unit testing, QA testing, end-to-end integration testing, deployment, and verification. The project must consider the lead time required to engage the services of these external teams as well as their existing availability when planning releases.

- **The impacted business and operational teams and their blackout dates**: There is nothing worse than having a well-planned project timeline disrupted because other business-critical parts of the organization have a scheduled blackout or brownout date on the deployment or go-live date. After impacted stakeholders and teams are identified, relevant individuals within the stakeholder teams should be contacted to determine whether there are major release schedule conflicts or blackout dates that could prove disruptive to the release process.

Planning for data migration and validation

Plan sufficient time and resources for data migration and data validation. Whenever possible, minimize the amount of current-state data that must be migrated to the target state. Data migration can be one of the most time-consuming and complex components of a transformation. Complexity is not only caused by the potential volume and quality of data to be migrated but also because of operational impacts that may occur during the migration process. When it comes to migrating data, simple is often better. The following is a list of important factors that should be considered in your migration strategy to de-risk your release.

Should you even migrate the data in the first place?

It may be an easy assumption to make that all data from existing systems will be migrated to the new ServiceNow system as part of a major implementation. However, there are multiple risks involved in legacy data that should be considered before deciding to migrate the data.

First, the quality of the legacy data for providing reporting and other business value insights should be considered. Data quality issues and a lack of good business data are major reasons for a large IT technology transformation and so a blanket migration strategy will often bring data of dubious insight dubious insight or quality into the target state may impact the subsequent value realization of the transformation.

Secondly, and possibly harder to recognize at the outset, is the complexity of transforming legacy data to conform to the target state. Major transformational implementations will typically lead to the adoption of new target state processes, procedures, and ways of working both inside and outside of a system. This change is likely to fundamentally change the definitions of business data between the current and target states. In a very simple example, if in the target state, incidents can automatically close if they have been resolved after 5 business days but in the current state, it takes 10 business days, then all target state metrics will automatically be impacted by the introduction of data from the current state. In more extreme examples, entire process steps or procedures may be altered, eliminated, or added in the target state, which can make the business data in the target state wholly incompatible with the legacy data from the current state. In these situations, data migration of legacy data should be carefully considered. At the very least, migrating the legacy data into an area separate from the target state data may reduce both its complexity and impact on future reports and metrics.

What parts of the data should be migrated and what strategies should be employed?

Sometimes, data migration is unavoidable – in these situations, it may still be worth considering what parts of the data should be migrated and where.

For operational data, try migrating to a *legacy* area or archiving the data. Operational data includes things such as open tickets, historical tickets, and active cases: data that is operated on by humans and viewed by humans as customer requests for services. There are usually two primary reasons why operational data should be migrated in the first place:

- **For compliance purposes**: Some operational data should be archived for several years. In these cases, strongly consider a migration plan that involves simply archiving the existing system's database for access in the future. Archiving tends to be much simpler from a release standpoint than migration, as no transformational logic needs to be defined and the archiving process can be done outside of any other deployment activities and so does not need to be on the critical path of the deployment activity.

  If archiving is not sufficient, for example, because the data needs to be more readily available, then consider migrating the data 'as-is' into a separate area than the target state data model. In this model, the current state data is essentially migrated into the target system in a 'like-for-like' manner and stored independently. Some linkages can be made between the legacy data and the target state data (for example, legacy incidents can be linked to target state users), but no transformation is done to make the legacy data conform to the format of the new data. This strategy reduces the complexity of data transformation logic and eliminates the impact of legacy data on the target state metrics.

- **For active operational needs**: Often, there is operational data that is still being actively worked on when the deployment to the new environment occurs. For example, what happens to open incidents or open cases when the existing system is transitioned to the new system? Even in this case, there are still options to explore before deciding that the data definitely has to be migrated. One option is to employ a sunsetting strategy where at the time of deployment, the existing system is not entirely shut down. For a certain sunsetting period, your teams will work in both the existing system and the new ServiceNow deployment, but with a very clear set of procedures to determine when to work in one over the other. For any operational data that remains open at the prescribed cutoff time (usually when deployment begins), these tickets will continue to be managed in the legacy system. After the cutoff time, new data will not be allowed to be generated in the legacy system (this is usually accomplished by disabling all the capabilities in the old system to create new operational records to prevent human error). In this way, the legacy system will have a finite number of existing tickets that still have to be worked on and closed. At the end of this period, the interim state in which both the current and target state systems are used can end, as the organization fully transitions to the target state.

When using a sunsetting strategy, there should be an estimate made on how long the interim period will last. One way of estimating this time is by measuring how many new tickets are opened in the current state and how quickly they are closed on average. This will determine the throughput of the teams closing new tickets and this throughput can then be used to anticipate the length of the interim period. For example, consider that 100 new tickets are opened per day and teams close on average 50 tickets per day. Then, the length of time needed for the cutover period will be the existing open ticket count at the time of cut-over, plus 100 tickets per day during the cut-over, divided by 50. In this case, if the cut-over itself takes 2 days – during this time, the legacy system will still be used - and there are 300 other tickets in the system when the cut-over begins, then you will need approximately 10 days to close all legacy tickets by the end of cut-over.

A final optimization with the sun-setting strategy is to preemptively close tickets below a certain priority and beyond a certain opened-by date. This strategy should be employed in situations where your legacy system contains an enormous number of *dead* tickets: tickets that for one reason or another have not reached closure, but nobody has worked on or looked at for a long period of time. For tickets of lower priority, these historical dead tickets may simply be ignored, as the relevant customers have likely already forgotten about them. This strategy could impact the customer experience, so should be employed only when the number of *dead* tickets is at a point where it is unreasonable to deal with them any other way. The existence of this situation is also likely a sign of underlying issues concerning team resourcing, lack of clarity about the type of service provided, customer education, and communications in the current state that should be addressed or improved in the target state implementation.

When all else fails, there will be situations where you absolutely must migrate operational data into the target state. In such cases, a release plan should account for a dedicated data migration team and the dedication of time to designing, planning, and executing the data migration itself. A common mistake made when planning for data migration is starting data migration too late or believing that it can simply be solved by the target state process design teams. Because data migration in this context will require a translation between the current to target state data and their respective definitions, the data migration team needs to be embedded within the target state design teams to fully understand how best to translate legacy data elements into the target state. . The complexity of this activity should not be underestimated, and so the data migration stream should be resourced as if designing net-new functionality, as opposed to simply moving data from point A to point B. This will also allow the migration activity to be started in parallel to other project activities, which will provide early visibility into any issues and mitigate the risk of last-minute complications.

Considering testing and training

The time spent on testing and training is an often neglected part of a typical release schedule. Even worse, when third-party vendors are placed under time pressure, this could be the first set of activities that is *replanned*, to the detriment of the long-term success of the project.

When considering the release plan of your transformation, anticipate the additional time needed to test the solution being created and to train impacted teams to adopt the changes ahead of time. The following are a few considerations to incorporate into your planning.

Budgeting time to create automated tests to aid future regression testing needs

When new functionality is developed, creating automated testing using ServiceNow's **Automated Testing Framework (ATF)** or any other automated test systems can significantly contribute to the organization's ability to be agile in later releases. A common misconception concerning automated testing is that it saves time immediately. In reality, however, automated testing adds effort to the implementation upfront, as creating comprehensive automated tests for features can be as complex as writing the features themselves in the most complex cases.

The value of automated testing is gained incrementally with every new net deployment of capabilities onto the platform. Over time, as capabilities increase, the time to fully regression test the platform to mitigate unforeseen impacts becomes untenable for even large teams. In these cases, time-to-value can be significantly impacted without automated testing, and any failures in the already released capabilities caused by changes in the immediate release can cause significant downtime and customer experience impact.

When budgeting the time spent on building automated testing, consider that automated tests tend to add at least 10 to 20% to the effort of implementation. For complex implementations with many changes to the out-of-the-box configuration, this number may be even higher.

Budgeting 20 to 30% of additional project implementation time for testing and training

Even with automated testing, you will need normal quality assurance processes for the functions that you have implemented as part of the release. To properly exercise the solution that has been created, testing should cover both straightforward and designed cases, as well as error cases and edge cases. Many different types of testing must be prepared, each with different levels of complexity and different degrees of time added to the project:

- **QA testing** is the most basic form of testing. The purpose of QA testing is to simply have testers run through the platform and check that all the functional capabilities in the target state design are working as described. QA testing should come with structured, detailed test scripts with clearly articulated expected outcomes. QA testing should be performed based on exercising the target state processes designed as part of the implementation. If a solution enables the execution of the target state processes as described, then it passes QA testing. However, if a described capability is missing, does not work as described, or does not fully enable the execution of the process as described in the target state design, then it is a defect. Even for the simplest projects, QA testing should be planned to take no less than 20% as long as the time spent on process design and technical implementation.

- **System integration testing** is the technical testing of technical integrations between systems. This term is sometimes used synonymously with QA testing but should be differentiated. System integration testing should be focused on making sure the APIs or interfaces between systems are working as described, responding as described, and sending and receiving data correctly. System integration testing should also test any performance requirements between integration interfaces if that is a part of the design specifications. System integration testing can be particularly complex because it requires coordination between teams and systems. Additionally, because each integration may perform a variety of tasks and have a variety of legal and illegal inputs, testing may not only require significant manual work but also work on building testing automation tools to help fully exercise the integration. Integration testing, therefore, is a more complex type of testing, and we recommend allocating 30% as much time as you spent designing and implementing integrations.

- **User acceptance testing** is the process of having the real business users of the system test the solution to identify whether there are any glaring issues with the design that could prevent the effective operation of the system. One common mistake made in large-scale ServiceNow transformation is for user acceptance testing to be the only time that actual business users see the system. As user acceptance testing is planned at the end of a build cycle and very shortly before training and deployment, identifying a critical defect or change in requirement at this time can be extremely disruptive to the project. Therefore, the previous testing cycles should have already involved business stakeholders, who were the primary drivers of the design of the product. Instead, user acceptance testing should be the opportunity to identify areas where the design may be confusing, identify incorrect usage, or discover edge cases in the design. The amount of user acceptance testing can vary greatly based on how many customer-facing components are in the target state design. Transformations with significant customer portal components should be subject to much more comprehensive user acceptance testing phases.

Environments for testing

Each of the test phases in the preceding section should be mapped to a specific ServiceNow environment. If possible, a different environment for each phase of testing will allow for greater structure. If you have fewer ServiceNow environments, consider grouping system integration and user acceptance testing into one test environment and executing other activities in a dev or QA environment. What is critical is that you are clear on what activities should occur in each environment to assist with logging and resolving issues.

Beginning training after substantial testing has been completed

A common release planning mistake is to neglect the time spent on training the team or parallelizing the training process with the testing process. The latter situation is a common shortcut taken by third-party implementors when the request for a proposal or service places significant timeline pressure on the third party. Parallelizing testing and training comes down to risk management. The risk is that during testing, significant defects or changes are identified that invalidate any previous training. Multiple options exist to mitigate this risk, but one obvious and direct solution is to simply do training after the testing has been completed. Depending on the acceptable risk level, training may start after the first few rounds of testing have been completed, with the expectation that any further changes will be minor and will not drastically impact the training previously provided.

Training is a very broad term, but in terms of project planning, in particular when third-party vendors are engaged, it generally refers to classroom training, either in person or virtually, unless otherwise specified. In some cases, a *train-the-trainer* model may be adopted, where a specialized team is tasked with training a set of operational resources within the organization (team leads or managers) to then train the rest of the teams on the changes that they will need to adopt. Classroom training can be reused and repeated by recording the training for future reference or future use with other teams. In some cases, classroom training may be determined to be insufficient for the organization, and in such cases, time and resources should be allocated as part of the project to be able to create and deliver more specialized training options.

Specialized training options include self-directed e-learning courses built using professional e-learning delivery tools. In-system training and hints are built using either ServiceNow's guided tours functionality or a custom-built functionality for agents. For specialized training that requires development or custom content creation effort, the creation of the content can typically begin before all testing has been completed – however, a reasonable amount of time must be allocated to performing the final validation and editing of the content post-training to incorporate the changes introduced as part of the testing and remediation cycle.

Structuring the implementation team

A ServiceNow implementation project is a team effort and making sure you have the right team is a critical part of enabling your project to be successful. While the size of the project and the size of the organization will determine exactly how many members your team has, every successful ServiceNow implementation team should have at least one of the roles that we will detail later in this section.

In this section, we will highlight the various *functions* that make up the implementation team and then highlight the *leads* for each function. It is expected that as your project scale grows, each lead will go from an individual to a leader with several team members supporting them in the execution of their roles and responsibilities. Sometimes, your organization may have many functionally similar processes split across geographical boundaries or business boundaries. In such a situation, it is recommended that you also nominate *leads-of-leads*: someone with a high enough level of authority to make decisions and changes on behalf of the individual leads.

The focus on *leads* and not individual team members comes down to a single fact: the most time-consuming and complex aspect of a transformational implementation is being able to make business change decisions quickly. Configuring a tool to accommodate the differing needs of various stakeholders is generally far less complex to do but can quickly reduce the long-term value realization of the transformation by driving up the platform maintenance costs. Far too many projects with an initial vision of transforming the business fail to deliver on this promise due to the lack of commitment or support on the part of the team members to make business change decisions that take advantage of the platform's capabilities better. Having the right empowered leads supported by their team in place to make these decisions is a critical aspect of a successful transformation.

Examining major project functions

Project functions are a simple way of dividing up the responsibilities of a project team by the general types of decisions and insights that the team leads bring to the project. A typical transformation should be comprised of at least the following functions:

- **The project steering function**: The project steering function is responsible for setting the goals, guiding principles, and vision of the transformation itself. They are ultimately responsible for providing the entire implementation team with what is required to achieve these goals. In situations where project constraints clash with the stated goals, guiding principles, and vision, the project steering function is also responsible for either removing the constraint or adjusting the goals to be more in line with the constraints available to the project. This function generally comprises senior business representatives, as decisions made at this level of the project can impact not only the project itself but also other parts of the organization and organizational resources as a whole. The project steering function is led by the *project executive sponsor* and supported by the *project manager*.

- **The business transformation function**: This function is responsible for designing the target state business processes and implementing the business changes required to reach the target state. Business change can range from something as simple as procedural simplifications on a team-by-team basis to something as drastic as creating new roles, and responsibilities and adjusting the overall business operating model to better serve the needs of the target state. The business transformation function is also responsible for establishing the standards of organizational business data to support the technology transformation function. The business transformation function is comprised of multiple teams and team leads including *process owners, product owners, functional lead(s), solution architects*, and *organizational change management lead(s)*.

- **Technology transformation function**: This function is responsible for implementing the technologies, such as ServiceNow, that will enable the business transformation of the target state. While there is an interplay between the choice of technologies and the direction and design of the business transformation, most successful transformation initiatives with high-value realization are always led by the business transformation and not the technology. Nevertheless, the technology transformation function is important in helping the business transformational function achieve its goals. Additionally, for a prescriptive platform such as ServiceNow, the technology transformation function must help inform the business transformation function of the capabilities and limitations of the platform to better align the business transformation design to maximize these capabilities. The technology transformation function is comprised of multiple teams and team leads including the *platform owner*, *technical lead*, and *solution architect*.

Examining team leads and their skill sets

When identifying the leads for the teams in the aforementioned functions, carefully consider both the individual and their contribution to the team. An effective transformational team requires individuals with specific skill sets and experiences to be successful:

- **Project executive sponsors**: Often the executive sponsor of the transformation pick themselves by creating the vision and plan and convincing the organization to transform. With that said, the executive sponsor can still augment their abilities: by including the right individuals to support them as part of the steering committee or advisory panel. For large transformations, the executive sponsor should also be communicating and building relationships with leaders in parts of the business that may be impacted by the transformation initiative itself. It is important to keep in mind as the executive sponsor that the transformation initiative envisioned may not only impact other leaders indirectly but may also require the commitment of other leaders to change their way of working in service of the broader objective to be truly successful.

  Project executive sponsors should be clear on the purpose, priorities, and constraints of their transformation and be focused on helping the team deliver value despite these constraints. A project executive sponsor should also understand, at least broadly, what the transformation has to accomplish to deliver the value and objectives of the transformation.

- **Project managers**: There is plenty written about what makes a great project manager beyond the scope of this book, but from a ServiceNow transformation standpoint, a great project manager should take on the following:

  - Be detail-oriented and interested in the work being done

  - Have a level of functional understanding of the ServiceNow platform

  - Have experience in running large system integration and business transformation projects

  - Be able to build strong relationships and be well connected within the organization

Project managers keep track of and are aware of the needs and impediments of the team and work diligently to ensure those needs are met and impediments removed. A strong project manager can hold the project to its scope and its constraints and is unafraid to inform the steering committee and executive sponsor when constraints may be clashing with the vision and value objectives of the transformation.

- **Process owners**: These are leaders who are made accountable for the health of a process or practice within the organization. An important distinction between a process owner and the leader of a department or team in an organizational hierarchy is that the process owner has accountability for the entire process across the organization. For example, IT *incident management* is a practice that may involve infrastructure management teams, service desk teams across multiple countries, and network monitoring teams. Each team likely has a distinct leadership, way of working, operating model, and reporting structure, some or all of which may need to change as part of the transformation to improve the IT *incident management* process and achieve broader organizational outcomes.

  Process owners should therefore either be senior leaders able to affect these changes across these teams or be supported and empowered by a senior executive (e.g., the executive sponsor) to drive these changes. A process owner should be able to identify required changes to the organization's people, processes, procedures, and tools that can enable the process they own to achieve the value goals of the transformation, and be able to make decisions that appropriately balance the value realized versus the costs for, and impact on, the organization.

  Individuals nominated as process owners as part of the transformation should ideally not only be able to understand what it takes to "keep the lights on" but also have a keen understanding of which value metrics truly matter for the organization and what can influence those metrics positively. You will likely have multiple process owners, each with their own scope of processes that they are responsible for.

- **Product owners or functional leads**: Should be experienced resources with a deep understanding of the ServiceNow platform and the leading practices and frameworks in the industry. Depending on the specific scope and objectives of the transformation, functional leads may need specialized experience in product management and product development, or user experience design. Certification in the relevant ServiceNow processes may also be helpful for the product owners and functional leads.

  At a minimum, all product owners should have the ability to translate the unfiltered wants and needs of their customers (whether internal or external to the organization) into working target state process designs enabled by the ServiceNow platform. While product owners must focus on the "*what*" and not the "*how*" when it comes to designing solutions that will satisfy the requirements of their customers, exceptional product owners also have substantial experience with the functional capabilities of the ServiceNow platform. This platform experience can help a product owner design solutions that maximize the usage of out-of-the-box platform capabilities to more efficiently deliver the solution to the customer.

- **Solution architects**: These are individuals with substantial experience of designing and creating solutions on the ServiceNow platform (and integrations to other platforms) to solve highly complex solutions. Solution architects bridge the gap between the technologies available to the organization and their inherent strengths and limitations, and the designs of the product owners intended to solve a particular business problem.

  Solution architects should have a great amount of technical knowledge and also strong product development experience. Knowledge of the technology ecosystem of the organization is also a strong asset, as it can allow an architect to envision a solution to a particular design requirement that utilizes the organization's resources better.

  Solution architects also set standards in the transformation process: they have to create the designs for the elements of the transformation that are shared across the various process and practices of the transformation. Some elements of this shared architecture include the **Configuration Management Database (CMDB)**, organizational business data design (such as location, user profile, customer profile, and product catalog), and business logic design (such as customer support entitlement design).

- **Organizational change managers**: Identifies the impact on the people, process, and procedures of the organization based on the target state design that will be delivered as part of the transformation. Organizational change improves the adoption of a transformation and therefore its value realization, and the organizational change manager is responsible for determining the right approach and executing the right activities to improve that adoption.

  To achieve this goal, organizational change managers must be able to understand what the impact on the organization will be based on the ongoing transformation, a task that is made difficult by the fact that preparation is likely to occur while the design and implementation of the transformation are still in progress. The lead should therefore have some experience with transformations of this type (and therefore some functional knowledge of the capabilities of the ServiceNow platform) to most efficiently identify the possible impacts of organizational change.

  The organizational change manager and their team is also an expert in determining how people learn new skills and adopt new processes and has an exceptional understanding of the techniques that can change people's behaviors.

  They will be able to identify the individuals most impacted by the transformation to mitigate the cost of that impact through communication, training, and ongoing support, and prioritize these activities based on the anticipated resistance of the stakeholders to change and the importance of the stakeholders to the success of the transformation.

- **Platform owners**: A leader responsible for delivering value to the organization using the ServiceNow platform. The platform owner must not only add or enhance platform capabilities to support the organization's transformation needs but also balance these changes to the platform against the resources available to them to support long-term platform operations. These long-term operational activities include resolving platform support issues, fixing defects, upgrading the platform, managing demands for enhancements and projects on the platform, and managing and maintaining platform data.

  A platform owner need not be someone with a deep technical understanding of the platform but should be an operational leader who understands budgeting, operational processes and procedures, business analysis, production ownership, and functional design capabilities. A platform owner's greatest challenge is working with their team and the organization's leadership to create and manage a long-term roadmap for the platform that continuously generates value for the organization and enables value realization at the organization by supporting initiatives that may not directly involve ServiceNow. Doing this job well requires a clear understanding of the available operational resources and constraints on the platform team, understanding the high-level capabilities of the platform, hiring and managing a team with technical and functional capabilities in support of the platform, and managing a pipeline of business demands and delivering on those demands with consideration of the priorities of the organization and the resource constraints of the team.

- **Technical leads**: A platform owner should be supported by a technical lead who is in charge of the configuration and technical maintenance and management of the platform. The technical lead of a platform should be deeply familiar with the technology of the platform and run a team that can effectively deliver changes to the platform to meet the business requirements and demands approved for it. The technical lead is responsible for setting configuration standards and for assessing the risk of change to the platform. The technical lead is also responsible for leading their team into implementing changes on the platform in a way that best balances maintainability, scalability, performance, and the alignment of the platform's foundational abilities with other dimensions.

Can't I just hire a consultant for all this?

Many ServiceNow implementations and business transformations involve third-party consultants. In the ideal situation, consultants bring expertise that cannot easily be found or maintained by the organization to enable and accelerate the transformation.

Many of these aforementioned roles can be filled by external consultants who bring experience and best practices not readily available within the organization. Furthermore, external consultants can provide objective advice and insights that internal resources may not feel comfortable providing.

While external consultants can bring hard-to-find expertise to enable your transformation, the one thing they need your support on is being able to create the right constraints for the transformation aligned with the outcomes you are looking for. External consultants can only provide the best advice for your organization based on the constraints you have established. For example, if you ask them to take you across a river and only provide a small amount of wood and stones, they will likely recommend building a raft even if a bridge is ultimately a better long-term solution.

Therefore, if hiring external consultants for one or many of these kinds of roles, you should follow the principles laid out in the *Defining the scope of the project* section of this chapter to maximize the value that consultants will bring to your transformation.

When engaging consultants you will also want to be clear on whether you are looking for targeted roles to be filled or whether you are looking for the consulting organization to deliver an end-to-end transformation. These services vary drastically in scope, cost and supported outcomes so a clear understanding of what support you need from consultants will help you engage with them more effectively and successfully.

Now that you have the right team for your transformation, we need to add a final ingredient that can turn a good team into a great team: governance.

Governance

Governance represents the rules, norms, actions, and standards followed by the transformation team. Governance is important because it creates consistency within the transformation and removes ambiguity from the transformational initiative. Ambiguity can negatively impact value realization, as it can create misalignment between actions and expectations. The word governance can sometimes have a bad reputation in organizations because it may have been implemented in a heavy-handed manner in the past or was not followed well. A common, flawed reaction to a bad governance experience is to try to avoid implementing governance. This misguided attempt to avoid governance is often combined with the misuse of concepts such as *agile* or *lean*, interpreting these new practices to mean an avoidance of governance instead of simply another way to structure governance.

The truth is that governance is critically important within a project. It does not need to be convoluted and complex, but it does require clear enforcement for it to succeed.

Accelerating decision-making with a good governance model

From a ServiceNow transformation project standpoint, one of the most critical aspects of governance is to enable the transformation to make effective decisions quickly. When establishing the scope, the vision and guiding principles are great at placing the appropriate constraints and directions on the project – however, without governance these constraints are unlikely to be enforced or followed over the course of the project.

A critical aspect of good project governance is to clearly define how decisions are made and what the decisions that should be made are. The latter can often be overlooked, but formalizing the types of decisions that should involve governance can significantly reduce confusion within teams and accelerate the course of the project. During the project, many types of decisions may be encountered, but not all these decisions will require a formal governance model. Let's take an extreme example to illustrate the point: very few transformations would benefit from having a formal governance procedure to decide whether an individual can be excused from attending a meeting. In this case, leaving the decision to the individual is often more than good enough. Let's look at two major decision types, often encountered during a project, that should always be tackled using formal governance.

Decisions that conflict with the constraints or desired outcomes of the transformation

Whenever constraints clash, or when the desired outcomes of the transformation are impacted, there is nearly always a decision to be made that must involve the governance of the transformation. Some examples of constraints clashing during a transformation include the following:

- Expediting customer orders on a new service portal requires the supply chain team to begin viewing orders from ServiceNow, violating a scope assumption that only IT teams would be impacted by the transformation. A guiding principle of improving customer experience is now in conflict with a project scope constraint of not impacting a particular line of business during the transformation.

- The IT asset management team would like a custom UI widget with complex requirements to mass update asset contract data. Having the widget could reduce the time spent on managing IT asset data by over 5,000 hours a year, but the change is high-risk from a maintenance standpoint. The constraint related to avoiding complex changes on the platform now clashes with a change to improve the agent experience and reduce the operational costs of IT asset management.

- As part of an HR service management implementation, it was determined that an automated integration to two other HCM systems could result in the full automation of several high-volume HR service requests, bringing substantial savings to the organization. However, the integration itself could push the project timeline out by 8 months and increase project costs by 20%. In this case, the guiding principle of bringing automation and improved employee experience is directly in conflict with the project constraints of budget and time.

In each of these examples, there is no immediate solution that is correct. Each decision is a trade-off between two or more dimensions that are valuable to the organization. Whenever a situation of this kind occurs, governance should be engaged. Defining situations where project constraints clash as a trigger point for governance also helps eliminate decisions and situations that don't cause a constraint clash from having to engage a broader governing body. In the previous example, if the constraint of not impacting non-IT teams is removed and no other constraints are affected, the project team is then free to engage the supply chain team to make the appropriate changes.

Decisions that affect the specific outputs of the project

ServiceNow transformations generally have a set of changes on the platform as an output that configures the platform to fulfill the requirements of the transformation. The transformation is also likely to include changes to current state processes and procedures that must be implemented to make the best use of the platform. In each of these cases, the exact target state process or platform configuration must be decided to avoid back tracking later in the project. Multiple target states can achieve the same outcomes, and many aspects of the target state design are therefore subject to preferences and opinions. Given this, the target state process design, platform design, organizational change plan, and execution design should all be recorded as formal decisions and subject to governance.

When target state design decisions are determined to be subject to governance the decision volume must be considered. For even a moderately sized transformation project there are likely to be hundreds of such decisions. If only a single governing body at the highest level of the project is responsible for making all such decisions, it can quickly be overwhelmed and become a bottleneck to the project. It is therefore important to reduce the overhead of governance through tiered governance approaches, as we will detail in the next section.

Reducing the overhead of governance

Even when the scope of governance is clearly defined to eliminate frivolous decisions, there may still be an enormous quantity of decisions that need to be addressed through governance. A tiered governance model based on the impact of the decisions can help further improve the effectiveness of governance and minimize impacts on the agility of the project.

A tiered governance model is based on the fundamental principle of making decisions closest to the impacted area and involving the least number of decision-makers possible.

Whenever possible, the project should designate key decision-makers at the process and platform levels so individual leaders (such as the platform owners or process owners) can be empowered to decide the final target state for themselves. This is true for design decisions that impact a very specific area of the organization or a very specific process. In these cases, it is almost always faster to simply let process owners decide on behalf of the organization how this target state will need to look.

A general rule of thumb for designating decision-makers and the types of decisions they can make unilaterally on behalf of the organization is to elect leaders that will be the closest to the impact of change (both positive and negative). The seniority of the decision-maker within the organization also matters for these types of decisions: junior leads may have a very narrow set of decisions that they are comfortable making, while a too-senior leader may be disconnected from the day-to-day realities.

While making decisions closest to the impacted area and minimizing the number of individuals that are needed to make a decision is generally preferable, transformations will often have at least a few large decisions that impact many individuals and decision-makers in other areas. In such situations, there needs to be a mechanism for governance to escalate decisions to higher decision-making levels.

Escalating decisions

Governance should create defined ways to escalate decisions and guidelines on when decisions should be escalated. Escalation is the process in which one decision-maker brings a decision they originally were accountable for making to a higher or different decision-making body. When governance is structured properly, the escalation process settles even the most complex decisions quickly by rapidly finding the right decision-maker for the decision. When governance is structured poorly, there could be many more decisions escalated than necessary, resulting in delays to the project as decisions pile up for higher-level committees to get together and sort out. An alternative problem is when important decisions are not escalated but instead stagnate, with a decision-maker both unwilling to make a decision but also not comfortable escalating it.

To prevent these problems, governance should formalize the situations where decisions should be escalated as far as possible so that there is little room for ambiguity or interpretation. Governance should also define the escalation procedure itself, along with where decisions should be escalated based on the type of decision. With all of these points combined, we present 10 rules to live by when establishing your transformational governance model:

1. Establish objective guiding principles and list out the objective project constraints to be incorporated into any decision.

2. A formal decision needs to be recorded whenever it involves conflicting constraints or guiding principles (e.g., a decision breaks one constraint but better follows a guiding principle).

3. A formal decision must be recorded to finalize the target state design of a process, procedure, or configuration. This decision can be a single sign-off against a suite of design choices to be more efficient.

4. All decisions must respect project constraints or explicitly authorize the violation of a constraint as part of the decision. This reduces the possibility of decision-makers making *toothless* decisions that are incompatible with the project or organizational constraints.

5. There should be at least one decision-maker designated for decisions involving the technical design of the platform. When in doubt, this should be the platform owner.

6. There should be at least one decision-maker designated for decisions involving the design of the target state processes. When in doubt, the process owner of the organization can usually fill this role.

7. When a decision by one decision-maker negatively impacts the interests of another stakeholder who is a decision-maker in their own right, the impacted stakeholder must be consulted.

8. If the impacted stakeholder does not agree with the decision and can show a violation of constraints or principles, the decision must then be escalated to a higher-level authority.

9. The operating committee is a higher-level authority involving all decision-makers at the process and platform levels and is led by the executive sponsor. Escalated decisions that only impact the stakeholders within the scope of the project should be discussed and made here. The executive sponsor makes these decisions based on a prioritization of the project's vision, guiding principles, and constraints.

10. If the decision impacts parties outside of the immediate project scope, the steering committee is a construct involving senior-level decision-makers, selected depending on the impact of the decision to resolve conflicts and impacts that expand beyond the scope of the project. The steering committee can make final decisions by voting or by engaging an individual of organizational authority a level higher than the impacted stakeholders and the executive sponsor.

Value management

With the right scope, considerations for the timeline, a release strategy for the implementation, a skilled team, and a clear governance structure to make them effective, your transformation should be well on its way to success.

Value management is the final aspect of the transformation, something that should be made an integral part of any scope, release strategy, governance model, and any decisions made. The underlying intent of value management is to optimize the transformational activities so that its outputs (business changes, process changes, and technology changes) are well aligned with the value measures that kickstarted the transformation in the first place.

In a simple sense, value management comes down to continuously asking the question, "Is this bringing value to the organization and what the project cares about?" The difficult part of value management is having an answer to that question for each decision made within the project.

If you have followed this book up to this point during the planning of a transformation, you should have already created layers of constraints and guidelines in the form of value statements, a clear scope, guiding principles, a project vision, and KPIs, which help the transformation steer itself in the direction of value realization. The following set of tips injected throughout the previously covered aspects of the project execution should augment every aspect of the project execution with the right value-management principles to deliver a better, more value-aligned result:

- **Establish quantitative measure whenever possible, each with a baseline and a relationship to a clearly defined business outcome:**

    Quantitative measures are generally better than subjective measures because improvement can be directly tracked. *Customer experience*, for example, can be a very ill-defined and subjective term, but "reducing the time spent on the portal before submitting a request" or "improving clicks before submission of a service request" are goals tied to quantitative measures that are both relevant to that particular business outcome. Improving these quantitative measures compared to the baselines can then be directly fed into the scope of the project, as detailed in the earlier sections of this chapter, to accelerate decision-making around target state design.

- **Subjective measurements can still be made quantitative through the application of frameworks and systems:**

  Some decisions will always be subjective. Which color is better? Are clear skies better than cloudy skies? Ask several people and you're likely to obtain many different answers. Similarly, in the case of a transformative ServiceNow project, many common transformational objectives such as customer experience can have highly subjective decisions associated with them. One common subjective dimension may be in the styling or layout of the customer service portal: is a top menu better than a side menu? Should an option be hidden in a menu item or be part of the menu header? Purely subjective decisions such as these are difficult to manage during a transformation because they can often result in decision-making stalemates.

  But by applying the right framework and using the governance model to come to agreements on what frameworks are used, it is possible to turn many such subjective decisions into more objective ones. In the simplest form, a voting system can turn subjective into objective, but more involved systems include card sorting, using A/B testing, or instrumenting the portal with action tracking, which brings more involved quantitative measures to these normally subjective systems.

  The key consideration to note is that there is almost always a way of breaking down a subjective dimension of measure into a more objective one, given the appropriate amount of time spent determining exactly which aspects of that measurement the organization truly cares about and how much investment it is willing to make to find the appropriate quantitative measure. When all else fails, utilizing a voting system and choosing an appropriate panel of stakeholders can be the final fallback strategy for tough decisions that must be made quickly.

- **When making target state design decisions, have decision-makers always ask the question, "How does this change contribute to improving our KPIs?":**

  With the right quantitative KPIs established, your baselines measured, and these KPIs tied to the desired business outcomes, the final aspect of integrating value management into the project is by tying all decisions to value, as well as considering the constraints mentioned in the previous sections of this chapter. Decisions that consider both value and constraints are the only types of decisions that matter for the transformation, as missing any dimension will likely result in an incomplete decision or a decision that is ultimately not impactful.

Summary

In this chapter, we have provided an overview of all the critical aspects of what makes a project successful. In summary, a successful transformation is one with a clearly defined scope that knows what business value it is trying to obtain, how to measure that business value, and has a clear baseline from which to improve. A team with the right skill sets and experience contributes to success, especially when organized using a clearly defined governance model that accelerates the decision-making process of the transformation by reducing ambiguity and subjectivity. Now that you have all the elements to make your transformation successful, it's time to look at some of the more tactical aspects of the transformation that you absolutely can't do without.

Part 2 –
The Checklist

Planning and value are critical, but so is taking care of our foundation. This section will cover the areas you can't compromise on because they form the basis for secure and maintainable ServiceNow systems.

This part of the book comprises the following chapters:

- *Chapter 5, Securing Your ServiceNow Instances*
- *Chapter 6, Managing Multiple ServiceNow Instances*
- *Chapter 7, Designing Effective Processes at Scale*
- *Chapter 8, Platform Team Processes, Standards, and Techniques*

5
Securing Your ServiceNow Instances

As you implement **ServiceNow** for your organization, you'll focus on delivering exceptional value, but all that value can be erased in the blink of an eye if appropriate steps are not taken to secure your instance. A single cybersecurity incident can erode confidence, expose your company to massive liability, and have a profoundly negative impact on all those involved. Fortunately, ServiceNow provides effective tools for its customers to improve its own security position, and this chapter will help you understand which tools are available to you and how to make use of them as well as instill a culture of secure practices in your project team.

In this chapter, we're going to walk through the following topics in order to give you a better understanding of what should be done to protect your investment in ServiceNow from malicious outside actors:

- The case for investment in security
- Planning for instance hardening
- Improving instance security posture
- Encryption
- Integration security

The only system which is truly secure is one which is switched off and unplugged, locked in a titanium lined safe, buried in a concrete bunker, and is surrounded by nerve gas and very highly paid armed guards. Even then, I wouldn't stake my life on it. (Gene Spafford)

The case for investment in security

The ServiceNow platform spans many critical use cases in the typical enterprise, from **human resources** (**HR**) to **information technology** (**IT**), from legal to customer service. The value that stems from having ServiceNow support for these processes is often rooted in the availability of relevant data and orchestration abilities, which means that any breach of the data in ServiceNow can result in the loss of information and the potential for a malicious actor to exercise control over systems or infrastructure in your network.

Unfortunately, the investment in the security of your ServiceNow instance should go almost unnoticed if all goes well. A strong security posture deters attackers who prefer to aim at easier targets, and years of uneventful operation can make it appear as if the need to focus efforts on security is limited. Nonetheless, the potential impact of even a single mistake is so large that you should be continuously monitoring and improving your ServiceNow security, particularly where your ServiceNow instance contains tempting targets for attackers. As a project leader or sponsor, you have the chance—and even obligation—to make a clear and effective case for securing your instances. This section will guide you through the arguments in favor of investment in security.

What is at risk?

The specific modules used in your ServiceNow instance will determine what information or access is at risk. In considering and justifying investment in ServiceNow, it is important to be clear on what it is that you're protecting. The following list will indicate specific examples from each of ServiceNow's major workflow areas.

IT workflows

ServiceNow has supported IT workflows since its inception and has come to rely on the ability to both store critical data about the detailed configuration of your technology systems and remotely execute scripts within your network. A fully developed IT workflow capability includes the following:

- Discovery processes with access to a large majority of your networked technology assets
- Lists of specific **Internet Protocol** (**IP**) addresses with details of the specific software (including versions) of those endpoints
- Incident and problem records, potentially including log data or sensitive data related to users
- Orchestration processes that can patch, reset passwords, or modify the configuration of your technology assets

Naturally, the number and nature of sensitive data elements vary between instances; however, these provide a useful starting point for considering the importance of securing your environments.

Customer workflows

With the expansion of ServiceNow's customer support capabilities, entirely new classes of data are being integrated into ServiceNow. Customer lists, contact information, and even pricing and sales-related data can all be part of a customer workflow deployment, and the increased visibility of the solution only magnifies the importance of protecting this information for the sake of your company and the customers who trust you. Customer workflow deployments include the following:

- **Personally identifiable information** (**PII**) about the end consumers of your products and services
- Competitively sensitive information about customer satisfaction levels subscriptions and entitlements
- Details of products and services sold to customers
- Details of appointments booked by your field service staff to resolve customer issues

A breach of customer data can hang over your company's reputation for years, making the case to secure your instances even more compelling.

Employee workflows

HR data is among the most sensitive information within an organization and is a frequent target of data breaches. While employee details in an IT-only implementation can be minimal, HR profiles are often populated not only with work contact information but also with home addresses and personal contact details. In an implementation involving employee workflows, the following data is particularly sensitive:

- HR-profile sensitive fields, including place of birth, marital status, and date of birth
- Details of HR cases, and—in particular—employee relations cases
- Integration credentials with access to HR systems such as Workday or SuccessFactors.

Creator workflows

The ServiceNow platform also offers nearly unlimited potential for customization to address new business needs. As such, the scope of potential data that can be included within creator workflows is nearly unlimited. Assessing the risk for creator workflows is best done on a case-by-case basis.

Securing leadership support for security

We hope the preceding section has impressed on you the potentially catastrophic ramifications of any breach of your ServiceNow instances' security. Given the impact such a breach could have, it is surprising that many implementation plans observed in industry do not have dedicated effort allocated to ensuring the security of the instances. As a leader or sponsor of a ServiceNow project, you have the right and obligation to ensure that appropriate actions are being taken to keep your company's data and systems safe.

The best time to raise instance security is early in the planning process of your implementation; the second-best time is *now*. When you present your case for allocating some of your company's scarce resources to an activity that does not have immediately visible outcomes, it's important to be neither dismissive nor alarmist regarding the risks.

In our experience, a clear articulation of the value and risk inherent in your instance—based on the factors laid out so far in this chapter—and the unique conditions of your industry and use case can be used to establish the need to actively manage security. The remainder of this chapter will help you determine what to do and when.

We recommend using this approach to present the *problem* of security together with the *solution* of a straightforward plan that can be accomplished with a small fraction of the efforts of your implementation team. If you present only the concern without a view on how it can be addressed, you will either be immediately asked to go and prepare the solution (in which case, you lose an opportunity to anticipate needs and be proactive) or in the worst case, you risk the problem being ignored. Let's get to work building your plan for instance security.

Planning for instance hardening

Knowing that you should be securing your instances but not knowing where to start is a tough place to be. Fortunately, with a few simple guidelines and tools provided by ServiceNow, you can make a big impact quickly. Let's start with a clear picture of where security should be applied in your ServiceNow deployment.

Implementation security measures

With security often seen as a "checklist" activity that must be addressed prior to go-live, it can be tempting to try to focus the security standards for your environments on the production instance only, hardening it as your initial cutover approaches. Teams taking this approach often restrict the data that flows to sub-production instances and adopt less strenuous controls in those instances. Others allow security-related tasks to be undertaken in a final hardening cycle prior to production rather than continuously throughout the implementation. Rather than adopt these approaches, you should consider securing all instances early.

Secure your instances when they are provisioned

You should be taking steps to secure your ServiceNow instances as soon as they are provisioned because a compromise to security at this stage (such as sharing initial administrative credentials over insecure channels) could immediately compromise all future work in that instance. Compounding this upfront risk is the default capability in ServiceNow to postpone administrative **multi-factor authentication** (MFA) set up for the first few logins, meaning that a simple username/password combination is enough to gain access.

Secure all instances that promote code to production

It's not enough to secure only your production instances because development, testing, and staging environments all impact the final code that makes its way to production. A compromised test instance,for example,will allow an attacker to insert a malicious payload into an update set or Application Repository package before it's installed on production. While a careful audit of every line of every update set would pick up on the issue, relying on such manual and time-consuming reviews to secure your data is a recipe for failure.

Planning for secure platform operations

Knowing that your instance will be operational for many years, it is imperative that you consider not only the day-one security posture but also the continuous operations of the instance over time. This means documenting the security standards that are applied and the decisions that were made in setting up the instance. The following sections will provide a more detailed review of how to leverage the tools ServiceNow gives you to secure your environments.

Improving instance security posture

Security is best thought of with a layered approach. You can always improve the security of a system, but the goal is to provide security that is just a bit better than "good enough," and a common and effective way to do this is with many redundant layers. One common example of this that many people are used to today is MFA. When you log in to your bank account, the bank likely sends you a security code to enter. This is an example of layered security because the probability of someone guessing both your password and the code is much lower than the chance of them guessing (or otherwise discovering) either one. The more overlapping layers of security you have in your environment, the harder it is for an attacker to penetrate and access your systems.

Fortunately, ServiceNow provides several tools that provide mutually reinforcing layers of security to address the needs of even the most security-conscious customers. The most common of these tools are discussed in this section.

Instance Security Center

After securing your instance admin accounts, the Instance Security Center should be your next stop. You can access Instance Security Center on the menu under **System Security** or at `http://yourinstance.service-now.com/isc`. This dashboard provides you with three important resources, as follows:

- Security metrics, including data export counts, login failures, external logins, and antivirus statistics
- Compliance scores that inspect your instances' security controls and provide aggregate scores for security and **Payment Card Industry** (**PCI**) compliance
- Security resources and **frequently asked questions** (**FAQs**) to assist in your security setup

The following screenshot provides an illustration of where to find the **Instance Security Center**:

Figure 5.1 – The Instance Security Center is found under System Security on the menu

In general, the Instance Security Center is one of the most underutilized resources on the ServiceNow platform because of how easy it makes it to drastically improve your instances' security. The following screenshot shows some of its features:

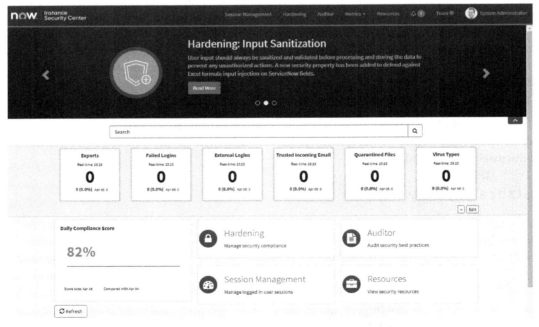

Figure 5.2 – ServiceNow's Instance Security Center provides a
suite of tools to improve your instances' security

The best place to start is with the security compliance metric score. While the exact number varies with the version, you can expect a score near 80% for a completely out-of-the-box instance, but with a few simple configurations, you can improve this score to at least 90%. Higher scores are better, but some security settings have an impact on **user experience** (**UX**) or functionality. This also means that configuring these settings early avoids you having to take away features from users later to satisfy a security audit.

ServiceNow provides multiple groups of security settings, each with **High**, **Medium**, and **Low** requirement levels. It is good practice to review all security settings that are marked as non-compliant and for each one to make a decision as to whether you can accept the ServiceNow recommendation to move them to a compliant state.

> **Important note – IP access controls**
>
> One setting that is very powerful is the ability to restrict access to IP addresses and ranges. These settings can be found in the **Low** section of the **Access Control** section. Many customers leave this feature off after go-live in order to make mobile and remote use easier, but it can be particularly useful in the early stages of an instance's configuration as it allows you to restrict access to only your company's corporate network. This significantly reduces the potential for an outside party to gain access over the internet while you are setting up other controls.

The Instance Security Center also provides the ability to track trends in login and antivirus activity. For basic deployments, the use of the security center can be enough to keep an eye on these metrics, but most sophisticated customers choose to tie these into their security monitoring to detect anomalous patterns early.

With the clear interface of the Instance Security Center and the built-in best practice check, ServiceNow has made the first steps of securing your instance the most straightforward—don't overlook it. While these security settings address the needs of many customers, ServiceNow offers further options to secure data, including options for various forms of encryption, which we will cover in the next section.

Encryption

A major topic in almost every large ServiceNow deployment is the selection of an encryption technology, and we'll dedicate some time to discussing how encryption should be addressed and making these options clear. We will also illustrate cases where each type of encryption might be used. Note that the most recent encryption updates are always available in the *ServiceNow Encryption Whitepaper*, available on the ServiceNow website or from your ServiceNow account executive. One of the most important things to realize is that in the ServiceNow ecosystem, ServiceNow has committed to addressing the in-transit encryption of data but leaves the decision of encryption at rest entirely to the customer.

How to approach encryption

Before diving into the selection of a particular encryption technology, it can be useful to consider your motivations and objectives for applying encryption. In almost all cases, the company implementing ServiceNow will have an information security policy or standard that classifies various types of data or various system use cases and then prescribes minimum technical controls for each classification.

Once you know the security requirements for the relevant classifications, you'll need to determine which of the encryption options provided by ServiceNow addresses the requirements while presenting tolerable trade-offs in cost and functionality. These classifications and the associated requirements should be discussed early, and a plan should be documented to avoid a scramble late in the project to adjust compliance and potentially introduce breaking changes from encryption.

Encryption types

ServiceNow offers a variety of encryption options that can be used in conjunction with other controls to secure the data in your instances. It's important to note that encryption alone is never sufficient protection for data in ServiceNow and must be combined with the recommendations of the Instance Security Center and general security good practices to secure your instances. We will cover the practical aspects of each, and for details on the cryptographic properties of each, we recommend referring to the *ServiceNow Encryption Whitepaper*.

Column-level encryption

ServiceNow offers the ability to encrypt specific attachments as well as targeted string, date, time, and **Uniform Resource Locator** (URL) fields directly on the platform with a feature set known as **Column Level Encryption Enterprise** or **CLEE**. This encryption solution allows administrators to set up role-based or application-scoped-based access to the cryptographic modules required to access certain data fields. CLEE is most effective when a limited set of attributes must be secured and where the encrypted data should be accessible to only a subset of the instance's users or applications. CLEE can be thought of as a complement to the **access-control list** (**ACL**) features that ordinarily provide **role-based access control** (**RBAC**) in ServiceNow.

Edge Encryption

The **Edge Encryption** capability in ServiceNow seems at first glance to offer a very similar capability to **column encryption**. Both methods allow specific fields to be encrypted while the rest of the data is unaffected; however, there are some important differences that determine when each method is appropriate. While column encryption functions fully within the ServiceNow cloud, Edge Encryption requires you to deploy an encryption gateway (or proxy) within your own network that allows you to encrypt and decrypt data for a specific set of fields inside your own firewalls, meaning that the unencrypted data never even leaves your network. This approach provides a very high degree of protection for specific fields but comes with some important trade-offs, as outlined here:

1. The setup and maintenance of **Edge Encryption** will require appropriate skills to deploy and maintain.

2. Unencrypted data is never present on the ServiceNow servers, which means that server-side logic and validations of the data are not possible.

3. Sorting and filtering of data in the **user interface** (**UI**) are impacted so that only exact matches or numeric comparisons can be made. You will not be able to use `contains` or other query types on edge-encrypted data.

4. Edge Encryption is not a useful solution for encrypting data that must be used within ServiceNow, and it cannot be used to encrypt all data in an instance.

Edge Encryption does, however, offer the most flexible toolkit for addressing stringent requirements such as data residency requirements or policies that require data to remain only within the customer network. Although Edge Encryption is arguably the most powerful of the encryption options, it is most certainly not right for everyone. A notable drawback is the inability to encrypt all data in the instance, which can sometimes be a requirement. Fortunately, ServiceNow offers other encryption options that can be used alone or in conjunction with edge and column encryption and that provide complementary protection.

Database Encryption

When security policy requires the encryption of all data or where the cost to determine which data needs to be encrypted would be prohibitive, it can be useful to adopt an encryption approach that simply encrypts all data. To enable this, ServiceNow provides a database-level encryption service called simply Database Encryption that causes the entire contents of the database to be encrypted at rest but without impacting application functionality (other than a typically insignificant performance impact). Database Encryption is often an efficient way to meet *encryption at rest* requirements, provided that the data can be encrypted in the ServiceNow data center and decrypted by any ServiceNow features that would normally have access to it (subject of course to ACLs).

It is noteworthy that Database Encryption can be combined with other forms of encryption such as Column Encryption or Edge Encryption to apply a higher level of control to specific fields while providing an at-rest encryption baseline for all data.

Full-disk encryption

In some small fraction of deployments, there may be a specific requirement to use self-encrypting drives, which provide protection against an attacker physically stealing the drives from the ServiceNow data center. This option would not typically be recommended unless there's a specific policy requirement dictating it as the capabilities and impact to the ServiceNow application are broadly similar to the database encryption features, which are generally less costly to implement and maintain.

While core instance security and data encryption are certainly good places to start, there is still one major area of the platform where security must be carefully considered: integrations.

Integration security

ServiceNow makes it incredibly easy to connect multiple production systems together, typically with web services that allow the real-time exchange of data between two applications on a network or on the internet. This ability for one system to remotely access and manipulate another introduces yet another place where security must be considered.

The security of an integration relies on the security of the source and target systems, the communication channel, and the authentication and authorization systems used. Each integration should be assessed for security on each of these elements, although special attention should be placed on authentication and authorization.

Source and target system security

When data is being transmitted in an integration, it is subject to the assumption that both sender and receiver endpoints of the integration can be trusted to handle that data securely. If either system is compromised, the data could be exposed. Data stored in two systems is inherently less secure from breaches than data stored only in a single system because all systems have some degree of security risk associated.

In a ServiceNow implementation, this requirement for endpoint security has two important implications, as follows:

1. You should ensure that the systems to which you send sensitive information comply with your organization's minimum standards for the relevant classification of data.

2. You should ensure that any customizations you might make do not decrease the security posture of the ServiceNow instance.

Securing endpoints

Developing integrations on ServiceNow has traditionally been difficult work, often requiring an experienced developer to work for days. While recent advances in Integration Hub have decreased the need for developers to custom-code integrations, you should be vigilant when an interface does require scripting. Utilizing Integration Hub can help to avoid common mistakes with security that can be made when scripting fully custom integrations.

An example of this endpoint security would be the use of the `GlideRecord` **application programming interface** (**API**) rather than `GlideRecordSecure`. In everyday development, it is common for developers to access data with standard APIs such as the `GlideRecord` API, which will query all data regardless of ACL constraints. In the right circumstances, this could allow an integration to be used to access data that was not intended to be shared.

Features such as Edge Encryption or CLEE would mitigate this risk if configured, but rather than rely on these additional controls, a developer should simply use a `GlideRecordSecure` call, which will evaluate ACLs within the context of the current users. This increases the configuration effort as appropriate ACLs will need to be set up (or roles granted).

It is also advisable to secure **REpresentational State Transfer** (**REST**) endpoints with defined ACLs to ensure that no endpoint is accessed by an account that has not been authorized. Note the two distinct but complementary objectives, as outlined here:

1. Prevent the intended consumer of an endpoint from accessing more data than they should have access to

2. Prevent an unintended consumer from being able to access an API they were not intended to access

If your team is not able to dedicate the necessary efforts or skills to securing endpoints, you should consider engaging a consulting specialist in integrations or utilizing only the interfaces supported by ServiceNow's Integration Hub.

Authentication and service accounts in ServiceNow

When an external system integrates with ServiceNow, that system will typically authenticate as a *service account*, also known as an *integration account*, which means that there is a record in the `sys_user` table that holds the identity of that integration connection. Managing these service accounts effectively is an important part of developing secure integrations and requires you to consider the level of access granted to each service account and the controls that will prevent unauthorized use of each account.

Recommendations for use of service accounts

Some general recommendations can be applied to the setup of service accounts in ServiceNow, as set out here. Following these will help you more effectively manage these important and sensitive credentials:

- **Grant least sufficient privileges**: A common principle in systems design is that of least sufficient privilege. The principle states that no account should have any greater level of access than the minimum that is sufficient for the account's purposes. It is far too often the case in systems integrations in general—including ServiceNow—that service accounts are provisioned with broad access through the reuse of out-of-the-box roles. This typically occurs when integrations are quickly set up and administrative or other elevated roles are granted to work around access errors. You may additionally restrict service accounts to interact only via web services by checking the **Webservice Access Only** box on the user account.

- **Utilize a common naming scheme**: Service accounts should be treated differently from regular user accounts as they are not tied to any one individual. To facilitate this distinction, it is often useful to name service accounts with a common prefix such as `svc-`.

- **Manage credentials securely**: The credentials of service accounts should only be stored in a secure credential store such as CyberArk. These credential stores are the corporate equivalent of password managers and can help prevent the loss or misuse of account credentials. If a secure credential store is not available, then you should not store the passwords in an insecure form. Rather, configure the connections, knowing that if the password is needed at a future date, it can be reset to a new value at that time.

- **Refresh credentials periodically**: Passwords should be refreshed periodically according to your company's security policy. If you have no policy on service account credentials, an annual rotation of the passwords should be considered.

- **Periodic service account reviews**: Throughout your implementation and during regular system operations, the roles of all service accounts should be routinely reviewed to ensure that the appropriate roles are assigned.

> **Important note – Integrations authenticating with end-user tokens**
>
> It is also possible in ServiceNow to set up an integration where end users generate their own tokens that allow an outside system to authenticate on their behalf. This is rare for production use cases in ServiceNow deployments and should be discussed with a qualified security professional if you believe it to be necessary.

Role of the ServiceNow MID Server

One of the most important tools in a ServiceNow integration specialist's toolkit is the **Management, Instrumentation, and Discovery (MID)** Server. The MID Server is a service that runs on a server or container inside the customer's network and brokers integration traffic by establishing a secure outbound channel from the customer network to the ServiceNow instance and using that channel to facilitate bi-directional integration.

MID Servers are typically set up in clusters, which can provide load balancing and redundancy for interfaces. When ServiceNow needs to integrate with resources on the customer network, the MID Servers allow it to do so without requiring firewall rules for inbound connections or complex proxy configurations.

When developing an integration between ServiceNow and an on-premises system, you should always consider MID Servers as these allow you to leverage ServiceNow's investment in a secure channel from the MID Server to your instance.

At the same time, consider the power that these MID Servers have in accessing your ServiceNow environment and ensure that you have secured them appropriately, taking special care to manage their service account credentials securely. In particular, MID Servers should always be set up according to the current installation documentation, which will protect the credentials and support a secure connection.

Summary

In this chapter, we have discussed the need for security in a ServiceNow implementation and helped you understand how you can build a case to allocate the limited resources of your project to ensure the systems and data under your control are protected.

We've covered the tools at your disposal to secure your instances through the Instance Security Center as well as the various data encryption options available on ServiceNow. Finally, we addressed the considerations for secure integrations within ServiceNow.

As discussed in this chapter, the concepts of security must be applied to all ServiceNow instances under your control. The next chapter will cover the management of multiple ServiceNow instances and provide you with tools to assess how to use your ServiceNow instances.

Managing Multiple ServiceNow Instances

Typically, ServiceNow customers will have multiple ServiceNow environments, also known as **instances**, which are independent copies of the ServiceNow application. In most cases, an order form for ServiceNow will include one production instance and one or more sub-production instances. The number of instances granted in an order is typically dependent on the size of the order with ServiceNow and, often, ranges from two to five total instances (although higher numbers are not uncommon among the largest customers).

As your ServiceNow implementation goes from vision to execution, one of the very first things the administration or development team will need to do is to set up your instances. Setting up these instances properly at the start will help improve the velocity of your project. Unfortunately, for many teams, this critical stage comes before the team has had enough hands-on experience with the platform and before they can be expected to design an optional environment plan. The goal of this chapter is to help you understand and plan for the setup of your instances, including the following:

- What are ServiceNow instances?
- Instance administration operations
- Code promotion and data clone flow
- Designing your ServiceNow landscape

What are ServiceNow instances?

Understanding the most effective way to buy, deploy, and manage your ServiceNow instances starts with a clear understanding of exactly what a ServiceNow instance is, which requires a little bit of knowledge of the component parts. In general, web applications consist of infrastructure running one or more web servers, application servers, and a database with more advanced applications having additional services or variations on this theme. Part of the value of a SaaS cloud offering is that the details of the servers and component parts aren't important to the customers, but it is still useful to think of a ServiceNow instance as `Core Application Code + Data` running on ServiceNow's cloud infrastructure (in the ServiceNow data centers). Note that we say *Core* application code because ServiceNow stores a lot of the JavaScript code that defines the business logic and even UI in the database.

An instance of ServiceNow can be seen from a few perspectives. Looking at it from a contract or infrastructure perspective, the *instance* you license is the space in the cloud needed to run a copy of the ServiceNow application. In this sense, you can say, "We bought three ServiceNow instances," which means, in this context, you are entitled to three environments according to your contract. This is important as different ServiceNow entitlements provide you with different underlying infrastructures; for example, production instances are often assigned more resources and might also be more isolated than development or test instances.

Typically, a ServiceNow instance has a specified role defined such as *Production, Development*, or one of the other roles we'll cover later in this chapter. In this sense, when you refer to your production instance, you would be referring to the cloud application and its associated database that your users and administrators can access. ServiceNow instances are often suffixed with their names, so a company named *Fake Example Company Ltd.* might have instances with the `fakeexampleco.service-now.com` URL for their production instance and the `fakeexamplecotest.service-now.com` URL for the test instance. This leads to the common convention where teams call their production instance *prod, production*, or by its full name, while other instances are called just by their suffix, that is, *test* in the preceding example. Typically, the major difference between instances is the content and use of the associated database with the production instance holding both the stable configuration and the actual day-to-day data used by your company, while a test instance might hold the latest prerelease configuration, a subset of production data, and any records that have been created/modified to validate functionality.

Finally, you might hear folks referring to the application version of ServiceNow as an instance and might say, "I need to try that in a Rome instance" or "I'm getting a Utah Dev instance," where *Rome* and *Utah* are the names of specific ServiceNow *family releases*.

If you bring these three views together, you'll understand that you can look at one of your instances and understand the infrastructure allotment, the instance role with the associated database, and the release of ServiceNow that is running. For the purpose of this book, we'll refer to the set of all instances your company uses as the **landscape**, although you might also hear people refer to it as an **environment** (which we avoid because an individual instance is also often called an environment in some contexts). With this foundation, we're ready to understand the basic administrative operations you can take within a ServiceNow instance.

Instance administration operations

ServiceNow provides multiple tools for its customers to use in managing their instances. This section will introduce two of these tools and their uses to set the stage for the workflows that you'll ultimately use to manage your landscape. First, we'll cover the zBoot operation followed by the system clone.

zBoot – a full reset of your instance

The most basic operation you can do to a ServiceNow instance is a zBoot (pronounced *Zee-Boot*), which allows you to reset your instance to the out-of-the-box configuration. A zBoot is a destructive operation in the sense that any work in the selected instance will be lost, whether that work is in the form of ticket data or configuration/code.

Uses of a zBoot

A zBoot is most commonly used in the restoration of a sandbox or development instance to an out-of-the-box state in preparation for some exploration or development activity. Note that once your company has active production instances, zBoots will become very rare, as you will more likely want your development instances to mirror production, a goal that can be achieved using the system clone operation, which we will cover in the following subsection.

Accessing the Now Support Automation Store

Like many administrative operations, you request a zBoot from the Now Support Automation Store. Now Support is the ServiceNow customer support portal found at `https://support.servicenow.com` and offers a broad catalog of request types, such as admin password resets, upgrade requests, and more. Your company's primary and secondary technical contacts will be provisioned with access to the Now Support portal, and they can add trusted administrators as additional contacts if needed:

Figure 6.1 – The Now Support menu

The Now Support Automation Store can be accessed from the Now Support header menu.

Requesting a zBoot

The zBoot option can be found on the second page of the **Instance Management** category, which can be accessed from the left-hand menu in **All automations** | **Service catalog** | **Instance Management**:

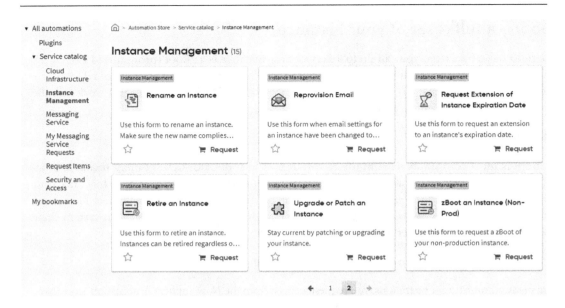

Figure 6.2 – The location of the zBoot catalog item in Now Support

Additionally, you can get to the zBoot catalog item and access numerous knowledge articles related to zBoots by searching for zBoot in the header menu search.

Selecting the zBoot options

When requesting a zBoot, the only parameters you need to specify are the instance name, the time for the zBoot to occur, and whether you'd like the instance to be populated with demo data. Note that due to the destructive nature of the zBoot operation, you cannot request a zBoot for a production instance online but will have to contact ServiceNow directly.

If you choose to include demo data but later want it to be removed, there is another instance management request item that is available to remove it. For this reason, demo data is often enabled in development and demo instances to avoid time-consuming data staging when showing the system's out-of-the-box functionality. Demo data should not be selected if you are preparing a production instance, as it can cause confusion during the setup process if you rely on records present in the demo data.

zBoot process

Once a zBoot request has been placed, a change record will be created, which you can use to monitor the process of the zBoot operation. Additionally, the change record permits the rescheduling or cancellation of the request, but you should be careful to only request a zBoot when you are ready for that instance to be fully wiped.

You can expect the process to kick off at the scheduled time and to take approximately 5 hours (although the actual time will vary). A zBoot operation clears all passwords and user profiles, so a new password for the system administrator (admin) user will be emailed once the operation is complete, after which you can use the instance.

System clone – copying the configuration and data from another instance

A system clone is likely the most frequent operation that you'll need to request, and it forms an essential component of the maintenance of a multi-instance landscape. Simply put, a system clone takes a copy of the *source instance* and duplicates it in the *target instance*, erasing any configuration and data in the target (with some exceptions that we'll cover later).

Uses of the system clone

System clones, or *clones* as they are often known, are used for three main purposes in the ServiceNow ecosystem. We'll cover them in order of their appearance in a customer's life cycle:

1. System clone go-live (optional)
2. Clones to synchronize non-production environments
3. Creating a production-like sandbox for upgrades, plugin activations, and more

System clone go-live

A go-live via the system clone is one of the options to update your production environment to match the development environment configuration. For many customers, it maximizes the certainty that your production instance will behave just like your test or development instances (depending on which you cloned from). Most critically, the cloning process will ensure that no configurations are missed. Additionally, it prevents common challenges due to system identifier mismatches, which occur when two records that look alike in different instances have different unique identifiers, a situation that can cause bugs in the configuration of ServiceNow. A clone go-live is also easy to execute as it is a mostly automated process once the necessary requests have been made and the change has been scheduled. This approach does have some drawbacks that should be considered as well.

A go-live with a system clone is straightforward, but in the simplicity of the approach, you do lose a degree of control and introduce some risks that alternative strategies can mitigate. The most compelling reason to take a different approach (for example the update set or application repository approaches) is the ability to leave behind unwanted configurations in the development instance and to more precisely control the configurations promoted to production.

Clones to synchronize non-production environments

Regardless of the go-live approach, you will eventually be faced with the task of using your sub-production instances to build, test, and deploy updates to your ServiceNow environment. This process is made exceptionally easy in the ServiceNow world using system clones, as you can replicate the production data and configuration exactly in your sub-production instance with special provisions to exclude or preserve data if needed. System clones are the best way to periodically refresh test and development environments and should be scheduled as an operational activity. Typically, a clone from production is appropriate after a major release of functionality, an upgrade to your ServiceNow instance, or when significant unplanned drift is observed between the production and sub-production data or configuration.

Synchronization clones can be scheduled on a reoccurring basis or manually. As a new customer, manual clones are preferred to avoid any surprises for developers, but with maturity and experience, a scheduled pattern helps to enforce good development practices and reduces the overall risk of deploying to production. Often, new ServiceNow customers find clones to be disruptive until they have appropriately tuned the process with data preservers and exclusion policies. We'll cover the function of these two important tools after addressing the final regular use case for system clones: the creation of a sandbox environment.

Creating a production-like sandbox for upgrades, plugin activations, and more

In the ServiceNow ecosystem, you will occasionally be faced with an attractive new piece of functionality such as a new and improved version of ServiceNow or an application or plugin that you would like to explore. In this situation, you will often wish to verify the functionality's compatibility and value alongside your existing configurations and data without incorporating it into your test or dev environments (which could impact test and development processes).

Cloning the production to an unused instance (or temporarily repurposing a non-essential instance) will allow you to explore the new functionality without disrupting regular operations. If you choose to take this approach on an instance normally used for a sandbox, training instance, or another designated role, you should plan to return it to the production configuration via a second clone once your investigation is complete.

Requesting a system clone

Unlike a zBoot, a system clone is requested and managed from within the source instance. You will find the system clone in the menu under **System Clone | Request Clone**:

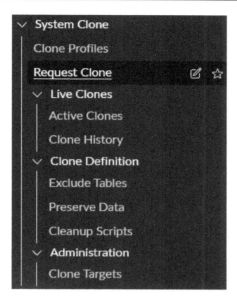

Figure 6.3 – The System Clone menu contains the tools for the management of clones

To schedule a clone, you can use the **Clone Targets** menu item to first specify a destination or target instance for the clone. Remember, you are requesting the clone in the source instance, so you will need to specify a target instance. When configuring a new target instance, you will need the full instance URL and an administrative username and password. This requirement helps to ensure that a third party cannot overwrite your instances. An additional step is required if the target instance is a production instance and involves setting the value of the system property: `glide.db.clone. allow_clone_target`. This property is a failsafe against the accidental overwrite of a production instance, so if you feel this is needed, consult the ServiceNow documentation and your ServiceNow support team. Typically this property is only modified if you are completing a clone for your initial go-live.

When scheduling a clone, you will need to select a clone profile, a target instance (from among those you've configured), a time, and one or more contact persons who will receive status updates via email. Additionally, there are several additional options available in the **Options** section of the form. These settings can be set in your clone profile and overridden on a per-clone basis. In general, it is good for your clone process to be as repeatable as possible. So, rather than choosing the settings for each clone, you should set these properties in a clone profile, which is then reused for each subsequent clone. If you are unsure about a specific setting, each label provides a link to the documentation explaining its function.

Clone profiles can also contain data preservers, exclusions, and cleanup scripts, which allow you to fine-tune the behavior of the clone. Typical use cases for these features include the preservation of user data, single sign-on parameters, and integration-related properties. The *Default System Clone Profile* is a great starting point for tuning your own clone profile and includes common exclusions, preservers, and scripts. Remember that your goal is for clones to be as non-disruptive as possible so that they can be run regularly. Time invested in a robust clone profile will be repaid with every successful clone, and when clone issues are observed, you should consider it a priority to update your clone profile to avoid the reoccurrence of the issue.

Now that you understand the basic operations used in establishing a ServiceNow landscape, we'll look at how configuration and code flows are promoted from your development instance to ultimately take effect in production.

Code promotion and data flow

The power of the ServiceNow platform comes not only from its out-of-the-box configuration but also from the ability to rapidly configure new business applications and processes. Remember that after your initial go-live, it is not possible to clone your configuration to production, as it would overwrite your data, so a different approach is required. ServiceNow provides multiple options for developers of these new capabilities, which allow them to move code from the development environment to testing and, later, production. There are two primary methods for the promotion of code between environments: update sets and the application repository (also known as *app repo*). In this section, we'll cover both and help you decide which is the most appropriate for your deployment.

Update sets

Update sets were the original mechanism for moving configurations between ServiceNow instances and remain the most popular and well-supported even today. The full use of update sets and their application is a subject addressed in the ServiceNow training and documentation and would consume multiple full chapters on their own. For our present goal of defining and managing a ServiceNow landscape, it is sufficient to understand them at a high level.

Essentially, an update set is a record of configurations made in one instance that can be loaded into another instance and then applied while checking for conflicts. Update sets are incremental and do not typically capture a full application configuration (although there can be some exceptions to this). The easiest way to think of update sets is as a recording of the work that happened in development, which is timestamped so that it can't accidentally overwrite a more recent update in another instance.

Working with update sets requires developers to pay attention to which update set they are using because all configurations made while an update set is active for a developer will be moved when that update set is loaded in another instance. As a developer, it's recommended that you keep a close eye on all the records in your update sets to ensure you don't accidentally bring along configurations unintentionally.

Update sets can be moved between instances as XML files or via an integration that ServiceNow provides and are reviewed for conflicts and issues before they are applied in a target instance. Some examples of conflicts can include a business rule that references a user group that does not exist in the target instance or an update to a record that has more recently been changed in your target instance. In both cases, you'll be notified of these situations when you are reviewing the update sets in your instance.

Update sets have many advantages and are uniquely well-suited for the promotion of small, modular updates of functionality to an existing production environment. Their size and the fact that they can contain as little as one single update make their behavior predictable when tested on similar instances. As you will appreciate, this also means that it is essential that your testing environments reflect the production instance because only the modified code is moved. This is okay in most situations, but in other cases—such as managing multiple production instances—it becomes very difficult to be sure your changes will have the intended effect. The alternative is the use of the ServiceNow application construct, which we'll cover next.

Scoped (and global) applications

ServiceNow has long had the ability to develop applications in contained *scopes* that help ensure that configurations are sandboxed in a way that they can be deployed with more predictable results in any ServiceNow instance. A scoped application, unlike a normal update set, captures the full current state of that application with the expectation that when it is applied, the entire application is updated to match the current code base. In this way, a scoped application deployment has a monotonically increasing version number, and when a new version is installed, it is expected to fully replace the old version.

The drawback of scoped applications is that you are constrained to a more limited set of APIs than a global application and that a particular application scope cannot contain edits to records from other applications. ServiceNow does allow the use of globally scoped applications through the use of the `sn_g_app_creator.allow_global` system property, but this is disabled by default in new instances, as scoped applications are much safer and less likely to cause challenges in deployment and maintenance.

When a scoped application is ready for deployment, it can be published to your company's application repository, which makes it available to install in other instances, such as your test and production instance. Also, you can export a complete update set, containing every record in your scoped application, which can be useful for moving configurations between different customer accounts, such as from one company to another.

Knowing both of these deployment options for the ServiceNow configurations and code along with the instance operations has set the foundation for us to start looking at practical advice for designing your ServiceNow landscape.

Designing your ServiceNow landscape

Both the most and least useful thing you can learn about designing a ServiceNow landscape is that the correct design depends on many factors, including your team's level of experience, the number of instances at your disposal, and the types of applications and configurations you'll be building. Fortunately, we'll look at more practical advice in this section, including the most common scenarios and principles you can use to inform your own design.

Two-instance landscapes

The easiest scenarios are when you have two or three instances with two instances being rare in all but the smallest deployments. In the case of two instances, you'll use one production instance as production, and the remaining instance will act as both your development and testing instance:

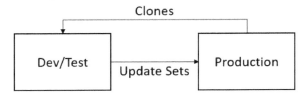

Figure 6.4 – Two-instance landscape design (using update sets)

Customers using only two instances almost always use update sets for code promotion rather than scoped applications, but there are advantages to using scoped applications too, such as the ability to work on those applications in a separate personal developer instance (a free instance that ServiceNow temporarily assigns to developers). We'll cover the considerations in more detail in the discussion on three-instance landscapes.

Three-instance landscapes

Most customers will have at least three instances, which provide the extra step of security because configurations from development can be tested in a more production-like test instance, which is free from other work-in-progress code. If you have three instances, you'll certainly want to set up a `Dev -> Test` , `Dev -> Production` code promotion flow while periodically refreshing your test and development instances with clones from production:

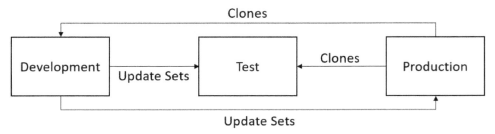

Figure 6.5 – Three-instance landscape design (using update sets)

A three-instance landscape can benefit a great deal from the use of scoped applications and the application repository because it allows you to easily compare the versions of code and configuration in development, testing, and production. If most of your configuration work can be completed in scoped applications, which is usually the case, then the use of the application repository will give you faster and more reliable deployments:

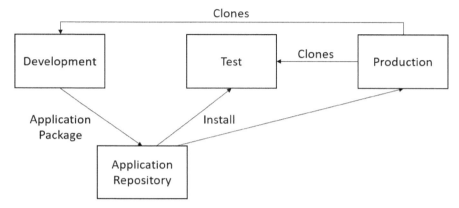

Figure 6.6 – Three-instance landscape design (using the application repository)

Additionally, it is possible to adopt a hybrid approach where a majority of configurations are promoted in the scoped applications but any minor updates that are required in the global scope are migrated either by enabling and installing globally scoped applications or by update sets.

Utilizing four or more instances

The natural next question is this: what about larger deployments of four or more instances? Let's look at each case and provide common uses for additional instances.

Additional quality assurance

Many companies, particularly those with strong internal control or regulatory oversight, require a separation of the instance where the development team conducts their testing from a **user acceptance testing** or **quality assurance** instance, where a separate team or group of users confirm the production-readiness of a deployment package. This creates a four-instance promotion chain that is identical to the preceding three-instance patterns except for the presence of an additional testing instance, often called **QA** or **UAT**.

Training

Where companies require frequent training in a production-like environment, a training instance might be used to reflect the current production functionality but with the freedom for users to create *testing* or *dummy* data during training classes. In most cases, training instances are only ever refreshed by clones from production and will rarely have code deployed directly to them.

Additional development instances

In some cases, multiple development teams need to operate in a single company but with some contradictory requirements. For example, an operational development team might be working on workflows for the current release, while a parallel team might be creating an entirely redone portal using a newer preview version of ServiceNow's next upgrade. In these cases, having two development instances at the same time can present a significant advantage in allowing work to proceed in parallel on different timescales.

Sandbox instances

Finally, a ServiceNow team might need to occasionally test things in an instance that is not otherwise used for operational activities. This might include an early preview of an upgrade, a new store application installation, or a development activity that is too risky to conduct in your development instances.

The composition of your ServiceNow landscape is determined by your business needs and obligations. Using these building blocks, you should be able to allocate the scarce resources of your instances to meet those needs and obligations.

Summary

With what you've learned in this chapter, you should be equipped to complete the basics of understanding, designing, and operating a ServiceNow landscape of two to five instances or more. You now understand the operations that are commonly used to synchronize and configure ServiceNow instances, along with the ways in which these operations can be used to create a deployment pipeline that takes code from development through to production.

For more detailed information on the management of scoped applications and update sets, you might also want to refer to the ServiceNow documentation and the system administrator and application developer courses. Additionally, it is a great idea to request a **Personal Developer Instance** (**PDI**) from the ServiceNow developer site (`https://developer.servicenow.com`) in order to get some hands-on experience with ServiceNow environments and the concepts of this chapter in practice.

In the next chapter, we'll look at how the processes you develop within these landscapes can be designed to deliver scalable value.

7

Designing Effective Processes at Scale

One of the most powerful capabilities of the ServiceNow platform is its ability to help an organization structure, standardize, and automate its processes. In this chapter, we provide an overview of the platform's automation capabilities and share some considerations around how to leverage them in a scalable and maintainable manner.

This chapter is divided into two major sections:

- An overview of ServiceNow's process automation capabilities and archetypes
- Designing process automation using platform building blocks and data

By the end of the chapter, you will be well equipped to spot opportunities where ServiceNow can be used to deliver organizational value and understand what you can do to estimate the complexity of the effort. You will also become familiar with the building blocks with which you can create the value-enabling process automation at your organization.

ServiceNow's process automation capabilities and archetypes

ServiceNow's process automation capabilities are underpinned by five major elements – integration APIs, service portal and digital forms, data-driven workflows, tasking and automated actions, and foundational shared data across the platform.

These elements work together to provide automation across processes. The typical flow follows this pattern:

1. A customer who wishes to engage in a particular process looks for that process's form in the *Employee Service Center*. An example would be *Obtain a new laptop* or *Offboard an employee*.

2. The customer is presented with a *digital form* and/or a checklist that they must fill out to start this process. This form may include error checking or logic that validates whether the customer is eligible to engage in the process and whether other options are available to them.

3. After the customer submits the form, the *workflow engine* uses the information provided by the customer via the digital form and underlying *foundational data* to create the appropriate *tasks* to be sent to fulfillers and/or performs automated actions to complete the process. During this time, fulfillers or customers may also be prompted to provide additional information, either through a task or through an email/chat.

4. The entire lifecycle of the process is tracked by the system, so the customer and agents have visibility of where the process is:

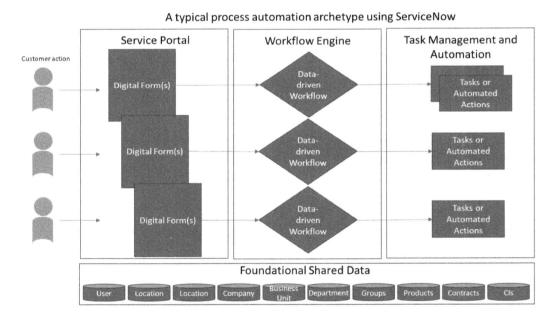

Figure 7.1 – A typical process automation design pattern on ServiceNow

Another common pattern of process automation for the ServiceNow platform involves an automated event or action captured by *integrations* that trigger the workflow engine to generate tasks or perform additional automation steps in response. This archetype is typical in event management use cases, such as when a monitoring solution detects an anomaly or actionable event and sends that information to the platform via integration. Once received, the platform can process the event, evaluate rules against the event, and then create manual tasks or take additional automated actions to resolve the event.

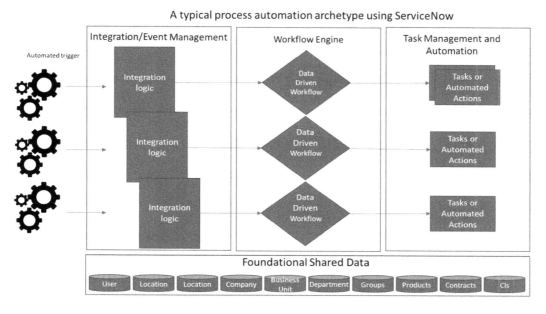

Figure 7.2 – The event management automation pattern in ServiceNow

The ServiceNow platform provides a number of different technical components to help you automate these two archetypical process automation patterns. Each technical component deals with one or more steps of the archetypical process automation flow. When designing process automation on the platform an important design choice is what technical components are the most fit-for-purpose to fulfill the requirements. Let's have a look now at the various technical components available to you and which part of the process automation pattern they service.

Technical components of the platform that deal with event intake or user input

For both the process automation patterns we have outlined previously, there's a *triggering event* or *action*, either from a user or a system. ServiceNow has a variety of ways of capturing these triggering actions. Let's look at the major ones.

The **Service Portal** and the **Service Catalog** are two major technical components that can be used to capture user input as triggering actions. The Service Portal is a customer-facing web portal that serves as a single stop or hub for a customer to be able to find the digital forms they need to trigger a particular process. The Service Catalog is where all digital forms are located and categorized in the catalog for easy searchability. The portal and the catalog work together as end users visit the portal, which contains the Service Catalog. The portal also provides search functionality to enable users to find catalog items easily without having to browse the individual catalog categories. The Service Catalog is a listing of digital forms, and ServiceNow has more than one technical component that provides these digital forms.

Record producers, **order guides**, **service requests**, and **HR service** are all technical components of the platform that enable the creation of digital forms. Record producers are the least specialized technical capability on the platform to produce digital forms. They allow for the creation of forms that can be presented to the end user and processed in whatever way the developer wishes, prior to the triggering of workflows. The record producer logic can even execute scripted code without the need for any workflows. The other technical components listed are more specialized features, intended to simplify or provide additional functionality for a specific domain or problem. In many cases, they are built on top of and take advantage of the record producer and order guide capabilities, adding additional functionality specific to its domain.

Service requests are built to provide a consumer shopping website experience. Service request forms are built within a framework that provides shopping carts, adjustable quantities, and pricing and price calculations. Service requests are at their best when used for the creation of digital forms that represent the ordering of physical goods, providing an online shopping experience out of the box.

The order guide may be used to create a *wizard*-like experience for the request process. An order guide is simply a digital form that can then trigger the addition of other digital forms as part of an order. For example, an order guide form for an employee to request a laptop may ask the employee in a single form about their peripheral requirements (for example, do you need a mouse, trackball, or a stylus?) and automatically generate a separate request for the peripherals instead of having the end user fill in each as a separate form.

The HR Service module is specific to the **HR Service Management** (**HRSM**) capability of the platform and augments the capability of record producers for HR-specific scenarios. The most important thing to keep in mind when considering the HR Service module in your design is that the digital form component of the module is simply a record producer. The HR Service module adds additional HR and HR-portal-specific configuration around a particular record producer identified as part of the HR service request to provide additional platform capabilities (such as lifecycle events).

You may also have the **Customer Service Management** (**CSM**) module, in which case there is also **case management** which are records designed specifically to capture various case types from external customers. Record producers are used as the forms to capture customer information to populate cases.

The information provided might feel overwhelming. Why does the platform have so many different ways of doing similar things? Some of it comes down to history and backward compatibility, but in other cases (such as the HR Service), it comes down to specialization of functionality. To simplify your choices, first identify broadly whether the process you wish to automate is an HR-specific process, a customer service process, or an IT process for employees, then pick the appropriate technical component (HR Service, case management and record producers, or service requests, respectively). Taking this approach, you will rarely make the wrong choice.

Web services are a platform capability that allows ServiceNow to provide REST or SOAP API endpoints that can be accessed by other systems. ServiceNow offers multiple ways to receive API calls, including the Table APIs, which enable other systems to perform CRUD operations on ServiceNow tables and scripted APIs that allow you to script complex behaviors upon receiving the inbound API calls. The `ImportSet` Table APIs allow web services to populate an import set table, which then runs corresponding transform maps to write the incoming data to a target table after undergoing a transformation process. Regardless of which web service options you choose, inbound API calls are the foundational platform capability that enables the triggering of workflows and actions from external systems and other platform capabilities. For example, **Event Management**, the ServiceNow module that is specialized in ingesting events from external monitoring tools and then triggering the creation of tasks via workflows or event rules, leverages web services for any push-based events coming from external systems.

The **Management, Instrumentation, and Discovery (MID) Server** is a Java application that runs in either a Windows or Unix environment on your local network. The MID Server enables ServiceNow to communicate with and execute commands and actions within your network environment. This platform capability can be used as a triggering action for various process scenarios, the most obvious out-of-the-box triggering actions being those provided by **Service Mapping and Discovery**. These capabilities utilize the MID Server to detect your infrastructure and services and monitor them for changes. A change in the environment detected by Service Mapping can then act as a trigger point for certain processes to engage, such as the creation of unauthorized change tickets.

Business rules and scheduled jobs can be used to trigger workflows based on changes or events occurring within ServiceNow. For example, a workflow may be triggered when an asset record's state is manually changed to `To Be Retired`. Business rules can detect data changes as they occur, while scheduled jobs may be used to scan the system periodically and then trigger actions appropriately.

Any of the preceding platform capabilities can be used to facilitate the capture of a triggering action or event, either from a user or system. As part of that capture, information may be collected; for example, a record producer or service request form may ask the user to provide information, such as what model laptop they want or whether they will be working outside of the office. The additional information collected, along with the triggering action, can then be used for the next step of the process automation pattern – workflows.

Technical components of the platform that manage workflows

Workflows are a series of pre-configured steps that execute automated actions in sequential (and occasionally parallel) order based on programmed rules. For example, a workflow may take a service portal form for a new laptop submitted by an employee, use the form submitted to determine where the closest stockroom to the employee is, and create a task and assign it to the IT service desk team at that stockroom to prepare the laptop for pickup by the employee.

ServiceNow has multiple technical components that provide workflow automation capabilities, the most obvious (but also deprecated) capability is the one aptly named **Workflow**. This was the primary platform capability to enable process automation until the New York release of ServiceNow, at which point it was superseded by a new framework for designing process automation called **Flow Designer**. As a platform designer in the Tokyo release or later, you should be aware of the existence of Workflows, but your designs should never specifically attempt to leverage its capabilities except in very specific scenarios. As of the Tokyo release of ServiceNow, there are still some areas of the platform that utilize Workflow instead of Flow Designer; one such area is defining a CI remediation as part of the CMDB Health platform capability.

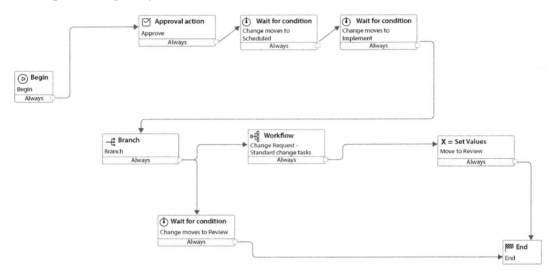

Figure 7.3 – The legacy Workflow capability of ServiceNow is still used in various areas of the platform

Flow Designer enables much of the same functional capability as Workflow but with a different interface, greater ability to debug and manage flows, and a modular design philosophy that encourages the reuse and composition of simpler components to create complex behaviors. The interface of Flow Designer, as seen in the next screenshot, is substantially different from the Workflow designer and follows a slightly different design philosophy. Still, its ability to launch a sequence of automated actions in response to a triggering event is comparable to workflows.

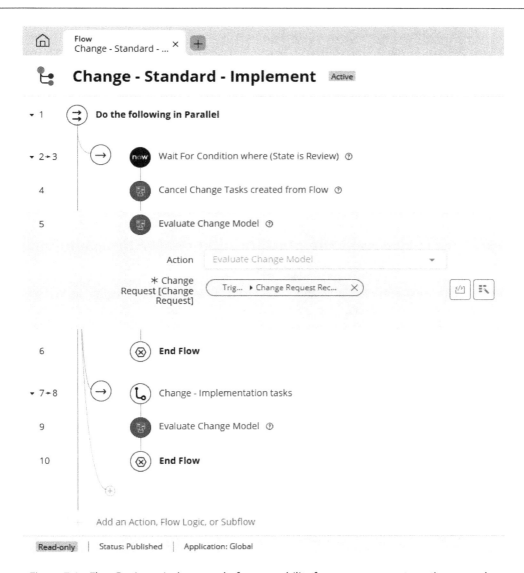

Figure 7.4 – Flow Designer is the new platform capability for any process automation scenarios

The **lifecycle events** capability in the HR Service Delivery application can be treated as a separate technical component built specifically for HR scenarios where a particular HR lifecycle event (for example, an employee is hired or transferred, or when health benefit renewals are imminent) may require a series of actions to be performed by employees and service fulfillers as part of the event. Lifecycle events are a powerful way to composite a variety of actions, forms, and steps together but were built specifically for HR use cases and therefore, limits their use except in that specific scenario.

Beyond these capabilities, ServiceNow offers the ability to use **Script Includes** to generate tasks and perform any action you desire, which can work as a bespoke workflow system, but as it is not purpose-built for this, we would not consider it a workflow capability comparable to the previously mentioned components.

From this point in the chapter, we will use *workflow* to refer to any of the aforementioned technical components. Workflows enable platform teams to manage and configure process automation in a structured way. They sequence one or more sets of actions performed by the platform in response to the triggering event and can pause before executing the next action in the sequence in response as it waits for user input or for additional triggering events to occur. Workflows provide the logic for when and what actions should be executed. The platform's task management and automation capabilities allow it to perform those tasks as orchestrated by the workflow components.

Technical components of the platform that deal with tasking and the execution of automated actions

ServiceNow completes the process automation loop by providing a tasking, assignment, and script execution engine that allows workflows to create and assign tasks to the appropriate individuals and allows scripts to trigger automated actions to occur.

Tasks underpin many capabilities in the platform where a manually actionable activity is assigned to a human being; this includes customer service cases, IT service requests, and HR service requests. Tasks have multiple things in common with one another, some of the most important ones from a process execution standpoint being the following:

- Tasks are assigned to individuals or groups, who can then work on the task according to their priorities.

- Tasks can contain descriptions that may include both generic instructions to agents regarding how to complete the task as well as information collected from the users/automated event trigger. For example, a task created because a monitoring system detected a server that does not yet have the latest security patch might contain details on the server that was detected and the results of the scan that was performed.

- Tasks track their *state*, which can be changed via automation (for example, in reaction to some action taken by an agent or user against the task) or manually (for example, an agent indicating that they have completed the instructions provided to them by a particular task). The change of state in a task can then be used to coordinate follow-on activities by the workflow. For example, a workflow responsible for an employee's onboarding may wait for a security compliance officer to mark a task in delivering workplace security policy training as complete before instructing the IT team to deliver laptops and access credentials to a newly hired employee.

ServiceNow also offers a specialized **Approvals** capability that, like tasks, is assigned to individuals or groups and has `state`. Like tasks, approvals can be used to coordinate process steps that are being automated by a workflow, such as when a workflow to send a purchase order to an IT hardware vendor waits for the approval of the finance team before proceeding.

Approvals differ from tasks in that they offer only a limited and specific set of actions against them – *approve* and *reject*. Approvals enable the capture of an approval or rejection reason from the approver when the action is taken and may be performed via the mobile application, email, or directly on the platform on a computer.

Tasks and approvals are ways for workflows to instruct humans to provide input to continue the execution of a process and scripts are a way for ServiceNow to trigger automatic actions as the process proceeds. Scripts can do anything on the platform – they can send an API call to an external system; update, create, and delete records; and even create tasks and approvals. Scripts in ServiceNow are packaged into different types of records, and the type of record determines when a script is executed and what information the script has available to it at the time of execution.

Script Includes are the most fundamental records for a script on the platform. A script included in a Script Include record is accessible by any other script in any other record in the system, and can be used to store re-usable code. For the purposes of process automation, one important record that may include scripts is **Flows**, including **Subflows** and **Actions**. These are Flow Designer components that can then be connected to support various process automation scenarios. Scripts inside Actions, Flows, and Subflows can perform automated actions in response to certain triggering events.

Workflows help determine the right actions to be performed at the right time, and tasks and actions enable those actions to be performed. The last step tying all the pieces together is foundational data, which helps workflows to execute at the right times for the right individuals, and tasks to be assigned to the right groups and provide the right contextual information to enable their completion.

Technical components of the platform that deal with shared and foundational data

Foundational data is the oil that lubricates the machinery of process automation. Many process automation scenarios of value require some type of foundational data to be on the platform, and in the best-case scenarios, great foundational data can improve reuse and reduce the need to build everything net-new in the face of new challenges and environments.

When we refer to foundational data, we specifically mean data that has the following attributes:

- It has a universally understood definition within the business

- The change (creation, update, deletion), storage (sources of truth), and usage of it is governed

- It is shared data that is the same data used by multiple business processes and means the same thing to each

From a ServiceNow standpoint, foundational data types are defined within the **Common Service Data Model (CSDM)**. These include `Location`, `User`, `Company`, `Contract`, `Product Model`, and others. It is important to understand that the CSDM and the foundational data types that ServiceNow can accommodate out of the box do not satisfy the aforementioned criteria for foundational data. For example, the CSDM clearly indicates that `Location` is a foundational data type and defines a default set of fields that can be stored within that record, but it does not provide guidance on whether locations should include the organization's buildings, rooms, desks, or parking lots. It is up to organizational and platform governance to establish the right framework for `Location` foundational data to be stored and used on the platform.

From a design standpoint, the important aspect to understand about foundational data is that ServiceNow is prescriptive about exactly what types of foundational data it formally supports through the CSDM. Any process automation design requiring data matching of those foundational data types should make sure that the data type is under governance, and if so, consume the foundational data in the appropriate way as prescribed by governance. Should the data required not be under governance (for example, your service request requires the location data of storage closets that are not currently stored in the **Locations** table on the platform), the designer should engage platform governance to introduce the data as foundational data on the platform. This process may be time-consuming but is necessary to avoid redundant or poorly maintained data on the platform and to encourage long-term platform health.

With well-managed foundational data within the system, many process automation use cases may begin leveraging this data to more efficiently enable capabilities that would never have been possible if data requirements were designed on a requirement-by-requirement basis. Imagine, for example, that foundational room location data for heavily utilized office spaces of an organization is managed on the ServiceNow platform. A service request that manages the booking of rooms could use the foundational room data to enable employees to pick the available rooms they want to book, while a smart office monitoring system could monitor AV equipment in rooms and dispatch IT to conduct repairs on equipment, reporting errors by providing the room information. More powerfully, because the room data is foundational data, repairs in progress could then be reflected in the room booking service request to prevent employees from booking rooms that have faulty AV equipment. This is only possible because the room foundational data is used in both processes and therefore enables each process to be aware of the other through that shared data structure.

Now our review of the platform's core capabilities to enable process automation is complete, it is time for us to put these capabilities into use, just as we did with foundational data, to create highly effective and reusable configurations.

Designing process automation by utilizing platform building blocks and data

With all the technical capabilities of the platform and an understanding of how process automation is composed using the major archetypical steps – *trigger events based on user or system input*, *sequence actions and tasks via workflow*, and *complete tasks and continue workflow*, you can now design process automation through these blocks.

The first step to designing process automation that works at scale is to know whether something is a good automation candidate. Thankfully, you should be able to use the archetypical steps outlined earlier in this chapter to identify them – those that follow the well-defined archetypical patterns are likely great candidates for automation via the platform, while those that don't may still be candidates but with a greater level of configuration complexity.

Let's look at an example. At a major international bank, a common and labor-intensive manual process involves the capture of emails from business banking customers requiring specialized services. Currently, customers submit their requests for over a dozen distinct service types via an online form that generates an email to a shared business banking service inbox. Once the email has been received, agents peruse the email to determine whether the customer is looking for one of the dozen services (for example, open a new account, obtain the last three years' transaction history, apply for a business banking credit card) that they are able to support. If the email is for a supported request, the agents will verify via a document stored on SharePoint what information to collect from the customer and which specialized support team's inbox the email should be forwarded to after all the information has been collected to fulfill the request.

Is the preceding example a good candidate for automation using the platform? We will use our archetypical steps to see whether it fits the pattern, by asking the following:

- Is there a clear triggering condition the platform can use to initiate the process automation? The answer to this is *yes*, as all the processes that can be engaged by the customer in this scenario start with the customer making the request themselves manually. This means that if we do use the platform to automate the request, we will likely leverage the service portal and a record producer or service request form to capture the customer request, or we might leverage the existing form the customer is using and send the information collected via integration to the platform. Either way, the trigger point is the customer clicking the **Submit** button on the form.

- Are there clear input data requirements that the workflow must consume to start sequencing actions and assigning tasks? In this case, the answer is again *yes*, as the agents currently have an existing list of specific requests that can be supported, as well as a list of specific questions to ask of the customer per request type. This data can therefore be structured in a digital form and captured up front prior to customer submission to save a valuable manual agent type and enable faster fulfillment of the request.

- Are there clear rules and logic defined in the process for the workflow to sequence actions and assign tasks based on the inputs? The answer is, of course, *yes*, as the agents have a well-defined way of determining which support team to ultimately send a request to based on rules captured in the SharePoint document. As long as the rules documented are deterministic (that is, the same inputs always result in the same behavior), they can be digitized and automated by the workflow so that humans no longer need to interpret and decide which team should be sent the fulfillment tasks.

- Are there requirements for foundational data, and is the required foundational data available? This is a tougher question to answer with the information available. If you are doing this design in the real world, you should immediately consider the foundational data elements that are likely involved in this scenario – customer information (account and contact) and organizational group information (assignment group) come directly to mind. There are also foundational data elements that may need to be established that are not yet available; for example, as you digitize each possible request for a service with its own intake form, you may find that some of the information you need to collect from the customer is better served when managed as foundational data. Some examples of this are branch information (location) and the customer's preferred contact information, pass-phrase, or other data that can help you authenticate the customer during support calls. As part of your design, you should try to produce a list of the data requirements and engage the platform architect to determine whether these requirements can be met using existing foundational data and what to do if some data requirements do not currently exist on the platform.

A final aspect of determining whether a process should be automated at all has nothing to do with the platform's capabilities; it's simply whether automating the process can generate value for the organization. As covered heavily in earlier chapters of this book, any configuration of the platform should be considered in terms of the value provided. In the preceding scenario, reducing the need for agents to manually route and obtain customer information may save agents one minute per request. This might not be worth it if the request volumes are small but may provide enormous value if the request volumes are very high. As an aside, product owners should also consider the value of increased service quality as opposed to just cost-cutting. A one-minute saving per request may mean greater service throughput overall and, therefore, greatly reduce service fulfillment times for each customer, resulting in value through better customer satisfaction.

Now, suppose we bumped into a potential process automation candidate that, in the end, failed to pass the preceding checklist. What might that look like, and how might we change the current state in such a way that we have better automation opportunities? The common failure scenarios for each of the preceding checklist items are listed as follows:

- Processes might fail the clear trigger condition check if the process itself is comprised of many subprocesses, some of which may be ill-defined. In such a situation, an immediate clear trigger condition may not be capturable because there are hundreds or even thousands of ways a process may be engaged. When you encounter such a situation, platform or process owners looking to automate should work with the process fulfillment teams to remove the ambiguity and/or to differentiate the various distinct processes to simplify the design of capturing the triggering event.

- When organizations have yet to mature their services to an extent where they are automatable, you are almost guaranteed to encounter at least one process where the required information to be collected from the service requestor is ill-defined. Sometimes this ambiguity has a compelling cause; for example, a service request to perform an engineering assessment on a building under construction may be complex enough that some back-and-forth data collection and understanding is required even if the obviously required data points are well-known. ServiceNow can support such complex scenarios where a case or request is comprised of a series of interactions and transactions, but it is certainly not specifically designed for such use cases.

- In the case of workflows and assignments, automation value may be lost or limited if the organization struggles to understand its own task flows. This can occur when a task assignment is heavily augmented by the agent's understanding of underlying team dynamics (for example, Tim has experience handling database issues while Stacy is a whiz at troubleshooting deployment scripts. Both work on the network team, but tickets from experienced service desk agents always find them and are resolved quickly, to the delight of customers). While this type of knowledge is not bad in and of itself, over-reliance on it to deliver great service can seriously impact the scalability of the organization. As teams become larger, training, separation, and specialization of capabilities and accountabilities become increasingly important to enable the team to scale efficiently. Additionally, these qualities also happen to benefit automation, as it enables patterns to be created (or inferred), which places the right cries for requests with the fulfillers and fulfillment teams with the right skills at the right time.

- Foundational data issues may prevent both the rapid scaling of automation technologies and the ability of automation technologies to deliver a positive return on investment in the first place. There are two common foundational data gaps for any large organization. In the first pattern, the foundational data of the kind needed to automate a specific process is available, but the data is of poor quality or difficult to access due to security or organizational restrictions or limitations. In this case, the focus should be on maturing the governance processes and investment around that data so it can be elevated to the level of quality to support the use case. In the second pattern, the data required may not be available at all. This may mean the data itself is entirely non-existent or that the data lacks the detail or specific information to enable the use case. Resolving the second scenario may be complicated by the reality that foundational data may be costly to maintain and that a single use case may not justify the introduction of a new foundational data type to be maintained by the organization. In such scenarios, platform and organizational governance must be engaged to establish the appropriate business case and design for the new foundational data. It may help to gather multiple use cases that could take advantage of the foundational data to form a stronger business case. The bottom line is that it will typically take collaboration, coordination, and funding across multiple business stakeholder groups to create the new foundational data and that the introduction of new foundational data to the organization is typically only tenable when multiple use cases can be identified and achieved together or as part of a roadmap to justify the overall investment.

Understanding the preceding failure scenarios should enable you to avoid implementation pitfalls ahead of any substantial commitment of resources. Treat the failure conditions as a set of risks for the project to be validated and mitigated up front to improve your chances of implementation success and of providing actual value in the attempt to automate your business process.

Now that we have the tools and checks to identify good automation candidates, let us discuss the considerations that can make the configuration performed by your platform team more maintainable and manageable, beginning first with the idea of creating common patterns.

Creating repeatable and reusable patterns to ease implementation complexity

The reuse of configuration is a tried-and-true technique to reduce the overall complexity and cost of existing code maintenance and new code builds. With code, the use of methods, functions, and objects enables multiple use cases to be achieved by reusing these components. Similarly, with workflows and forms, ServiceNow provides the capability for reusable components to be created, which can then be used in multiple scenarios. There are multiple advantages to this approach. First, smaller building blocks are easier to test. For example, it may be very complicated to test a ticket booking application end to end where a user submits a form, the form is sent to an agent to be validated, and then an API call is made to a vendor system to pay and then eventually receive a ticket number and unique digital ticket. The build and testing of this application would be vastly simplified if it were divided into a series of independent configuration components. One component may be a form that collects user data, one component deals with the validation of the form information and routing the form to the appropriate agent, another component manages the collection of payment information and the sending of payment information to the vendor system, and finally, a component processes the vendor system response and reacts appropriately.

ServiceNow not only enables this separation of configuration into multiple parts via individual scripts and functions, but it also enables this composition pattern in its workflow designer. The most important platform capability to know in the case of creating repeatable workflows is Flow Designer. Flow Designer offers Subflows and Actions, both of which are meant to enable platform developers to create repeatable components that can then be leveraged by multiple flows, achieving an ideal *build and test once, reuse many times* situation. The biggest difference between a Subflow and a Flow is the lack of a trigger condition. Subflows are triggered by Flows (or other Subflows) instead of being triggered due to a change to a record. Once a Subflow or Action has been created, new Flows can leverage these components, pass data into the Subflow or Action, and then collect the outputs of that Subflow or Action once they are complete.

For digital forms, ServiceNow's Service Request capability has two major components that can help with reuse. First, variable sets can be created, which package up a number of elements that can appear on a form and enable the behavior to be defined once and then reused on multiple forms. The most common example is a variable set created for any physical goods-based orders, which asks a user to input their preferred pickup location and preferred pickup time. The preferred location and preferred pickup time fields, along with help text and any form behavior, may be grouped into a variable set, which can then be added to any new form that involves the order and delivery of a physical product to save development time.

Order guides in ServiceNow provide a different way of reusing configuration: they provide a form that collects user input and then helps select and automatically fill in other forms, such as service requests.

Leveraging these capabilities, designers can identify appropriate functional components that can then be made reusable. Luckily, there are many such patterns that are universal across organizations and processes:

- **Approvals**: These are commonly repeated workflow patterns that can be made into a Subflow and then reused across multiple processes. Many processes, from ordering laptops to requesting a corporate credit card, may require approvals from a manager or senior executive before they can be fulfilled. The act of sending approvals to the right individuals and pausing the workflow until the approval has been granted is a common Subflow pattern that can be used across multiple processes.

- **Repeatable service request form elements**: As mentioned earlier, these are always great candidates for the use of variable sets. It is best to group these elements into variable sets when there is a clearly repeatable functional purpose, such as to prompt users for their preferred pickup location and time, to prompt the user on whether the request is being made on behalf of someone, and if so, to ask for information related to the intended recipient of the request. It is usually not recommended to group form fields that are unrelated to each other but commonly appear on the same form together. For example, even if the preferred first name, last name, and short description of the request frequently appear together, they may not be the best fields to be included in a single variable set, as the functional purpose of collecting first name and last name is different from the functional purpose of collecting a short description of the service requested. In such a case, a variable set can be made to collect the user's preferred name information with the preferred first and last name fields, while a short description can be kept in a separate variable set.

- **Repeatable sets of tasks**: These can commonly be created as Subflows. Many organizational processes may involve repeatedly performing the same series of activities. An example may be the shipping of a physical device to a particular office location. Regardless of the physical goods shipped, the mailroom needs to request and print a new shipping label, package the physical goods that are to be delivered, affix the shipping label to the package, and then bring the goods to be shipped to the shipping dock area of the mailroom. This series of actions may be created as a Subflow agnostic of the type of goods that needs to be shipped. Specific workflows for the shipping of goods can then reuse this Subflow and only provide additional differentiated steps for the preparation of the physical goods.

While repeatable components are a good start for scalable and maintainable design, there is a limitation – some repeatable components may differ in small ways that prevent repeatability. For example, while many processes may require a two-level approval, the individual who must do the approving is not the same for every request of the same type. An extreme example of not using data to drive this workflow would be to create a workflow for every unique approver. To avoid this issue, an additional design element must be leveraged for true reusability: that of data-driven components.

The concept behind data-driven repeatable components is to create repeatable patterns that act on specific input data. In the case of approvals, for example, instead of creating an approval Subflow for every person who may be able to approve a task, create an approval flow that sends the approval to a user or group that is designated as the *approver or approval group* for the user making the request. In this way, the variable input data of the user's approval group is provided separately from the repeatable behavior, enabling the creation of a repeatable approval Subflow that can send approvals to any user if the required approval data is provided as the Subflow input.

When creating your Flows, Subflows, scripts, or other components, consider what data elements being used in the design can be decoupled from the behavioral configuration and passed in after the fact to maximize the decoupling of your behavioral components from the underlying data sources.

Now that we have looked over some of the fundamental design principles of creating scalable and reusable designs let us summarize with a checklist of good software design principles that can be used to assess any design for its reusability and scalability:

- **Reusable components should be decoupled from concrete data sources**: If a Flow, Script, or piece of code requires specific data to function, the data should be passed to it as an argument or triggering object. This provides easy reusability when the data source changes, easier automated testing (through mock input data or mock data sources), and configuration understandability (by enabling developers to immediately see all the data needed by the component to function).

- **Reusable components should follow the single-responsibility principle (SRP)**: This is a computer programming principle coined by Robert C. Martin in the article *The Principles of OOD*, which states that "*A class should have only one reason to change.*" The concept behind this statement is more easily illustrated by an example. Suppose there was a Flow responsible for triggering the ordering of computer hardware (laptops and desktops) when new employees are hired. Such a Flow may be triggered by the hiring of an employee and may determine what hardware needs to be ordered by the employee's level and role. If built as a single Flow with no sub-components, then the Flow might need to be changed either because the event triggered by the hiring of an employee changes or because the rules for what hardware needs to be ordered for employee levels and roles change. In such a scenario, it is better to split the Flow into sub-components, one responsible for triggering the appropriate order process, and another that determines what hardware must be ordered according to the hired employee's role and level.

- **Reusable components should avoid being stateful**: Stateful means that a particular Flow, Script, or other component relies on knowing the state of a system to effectively do its job. Another way to think about stateful configuration is that the component displays different behavior based on what occurred before the component was engaged. In contrast, a stateless technical component does not care about or need any information beyond what is provided to it as its input to perform its function. Stateless components are a lot easier to test, as any data points needed by the component can easily be provided to it through mock test data, which also means that it is easy to automate the testing of different branches of logic by varying the test data as needed. In contrast, stateful components can be difficult to test as not only must input information be provided, but other dependent state information may also need to be set up prior to running the test scenario. Stateful components are also much harder to debug and for team members unfamiliar with the configuration to understand as they contain dependencies on many *hidden* variables that may not be obvious through casual inspection, therefore, impacting the scalability of the team.

Armed with the preceding principles, your team will be on its way to creating more reusable and scalable process components that can enable more rapid automation over time. Never forget that automation and creating better building blocks to enable faster automation is a continuous improvement process – regular code reviews, cleanup and refactoring of cold designs, and targeted re-engineering are needed over time as the team draws from previously learned lessons and as the patterns of business requirements and needs shift with the times.

Reusability, testability, and scalability are major software engineering disciplines about which thousands of books and articles have been created and written, most of them applicable to more than just ServiceNow. We strongly encourage your team to dive further into the topic as your capabilities mature and as you move to increasingly higher levels of automation.

Summary

In this chapter, we discussed how ServiceNow provides process automation capabilities to the organization through multiple technical platform components. These components are either specialized for specific purposes, such as HR Service Delivery, or are generalized and used as building blocks for process automation across use cases, such as Flow Designer.

Not all processes are great candidates for automation immediately, and before a team begins the process of automating a process using the platform, they should do preparatory work to identify and prepare foundational data and remove ambiguity in decision-making in the existing process so that it can be easily automated.

Finally, we discussed how thoughtful design and the use of technical platform components can improve reusability, allowing teams to quickly deliver new automation, leveraging existing pre-configured capability. The creation of reusable components decreases the risk and complexity of new configurations by reducing the number of tests required and by improving the ability of the team to test, which is critical to scaling up the automation capabilities of the organization.

The ability of your platform team to automate processes is only one of several crucial factors in achieving value from the platform. In the next chapter, we will look at how the governing and guiding processes of your organization can have a significant impact on value realization, and how to structure them for the best results.

8

Platform Team Processes, Standards, and Techniques

With your processes designed and your organization aligned, it's time for your team to roll up their sleeves and configure the platform to support those amazing processes in a user-friendly way.

So, where do you start? How does the platform team ensure what they are doing will not compromise the integrity of the platform? How do you minimize the work of maintaining the configuration going forward? I thought we were just going to go with **out-of-the-box** (**OOTB**) anyway?

This chapter will go into the details of how the platform team is run, cover contentious topics (such as the oft-brought-up *customization versus configuration* discussion), and go over the broad processes and technical patterns that can improve the maintainability of configuration on the ServiceNow platform for your organization.

Entire books have been written on how to run development teams, coding design patterns, architecture best practices, and coding coding conventions and so a single chapter will never be able to cover everything you need to know in detail. Instead, we will provide you with numerous building blocks across the entire life cycle of a platform team's work, from technical techniques to process improvement techniques, and show you where to do additional research to supplement the pointers provided in this chapter. We hope that you will be able to draw upon these building blocks to create a platform and team that will be envied by all. We will be jumping between process and technical advice as we go through this chapter, but broadly speaking, we will try to follow this order in terms of the themes of our advice

- Platform management and technical standards, including the techniques used to apply those standards

- Operational processes, continuous improvement strategies, and technical recommendations

Platform management and technical standards

At the risk of sounding like a broken record, management, governance, and standardization are frequently the missing ingredients for realizing long-term value in ServiceNow (or leveraging any other technology). When it comes to the technical integrity of the platform, standardization, management, and governance should consider multiple dimensions.

Setting standards to manage platform maintenance and risk

Any configuration change on the platform, from adjusting the font on the service portal to creating a scoped application to manage travel bookings, will increase the effort to maintain the platform going forward. More generally, any change in the configuration on the platform will bring risk to the platform in a variety of ways, from the risk of security breaches to the risk of impacting platform performance, to the risk of affecting the customer experience.

Because any change to the platform brings risk, any change on the platform must be balance against that risk. To balance risk, another dimension must be considered, and the typical one on the other side of the decision fulcrum is business value.

A change on the platform is only preferable if the business value it brings to the organization outweighs the risk it brings to the platform. However, *risk* and *business value* are both metrics that are subjective to the organization and must be established as part of platform management standards based on the profile of the operational team maintaining the platform. For example, a highly technical platform operational team might evaluate technically complex configurations as lower risk than a less technically skilled team.

Therefore, the terms *configuration* and *customization* are misleading as they falsely suggest that there's some universal property of a change that can easily be used to determine whether it is safe (configuration) or dangerous (customization). The reality is that there is no such thing as a safe change and you should instead consider if the risk of change is justifiable when considering the business value it brings.

Throughout their day-to-day activities, the platform team might need to evaluate hundreds of incoming requirements – all of which might require some level of change on the platform. If each one must be evaluated carefully by a committee, then the time-to-value will plummet. The evaluation standard that should be established must allow the platform team to quickly determine whether a change is worth doing and, ideally, help the team make decisions without having the involvement of senior leadership whenever possible.

One way to establish a standard to accomplish this is to create a table of the types of changes that can be made on the platform and assign a relative level of risk against it. Changes deemed *low risk* can be made at the discretion of the technical implementer, or the product owner, given that at least some level of business value is provided by the change. In comparison, changes that are deemed *high risk* might require higher levels of scrutiny by platform governance to justify (for example, they might require a business case, require review by an architect, or more).

The following sample table provides a solid starting point for your organization's personalized platform change risk assessment standard. You should work with your platform team to determine how to best categorize platform changes across a spectrum of risk and where to establish your threshold for "high risk." Also, you should be aware that this approach does still have limitations. For instance, it does not properly consider that a very large number of small, minimal-risk changes might still have a large impact on the maintainability of the platform. There is always going to be a trade-off between having a reference to accelerate decision-making and considering each decision carefully – it is up to the platform team to reach a model over time that balances the two for your organization:

Type of change	Recommended change assessment
<ul><li>Create a report or dashboard</li><li>Create a knowledge article</li><li>Add data to an OOTB data-driven functionality (SLAs and assignment groups)</li><li>Add a field or change a label on a record producer or catalog item</li><li>Change a form layout</li><li>Create or update a text-only notification</li><li>Make a field mandatory or hide a field using a UI action</li></ul>	Lower risk. Does not require a business case beyond product owner sponsorship.
<ul><li>Creating a new case type</li><li>Creating a new CMDB class</li><li>Add a notification with dynamic content</li><li>Creating a new workflow</li><li>Making an ACL change</li><li>Creating an integration into an external system</li><li>Creating a new scoped application (regardless of the actual makeup of the scoped application)</li><li>Creating a custom Discovery probe</li></ul>	Higher risk. Requires engagement of the business sponsor and platform architect to formally evaluate the balance of risk of maintenance and risk of change against the business outcome provided.

Table 8.1 - Assessing different kinds of change

Finally, now that all changes on the platform have been accepted as having some level of risk, the platform team should stick to the literal definition of OOTB to avoid confusion – a capability or functionality should only be considered OOTB if it requires no change from what is provided by the platform. That is, changing the ServiceNow logo for your company logo will not be considered OOTB, and changing the color of the agent side banner will also not be considered OOTB.

Standards and tips to manage platform changes

Now that we have standards on whether a change should be made on the platform or not, we can set standards on *how* changes should be made to the platform.

The platform team should establish a set of instructions that can be kept in a knowledge article on the platform for the team or as a document stored in a shared repository that can be referenced by every member of the team when platform configuration changes are being planned and made.

We recommend that the team considers, at the very least, standards and controls in the following areas.

Naming conventions of configuration records (business rules, UI actions, and script includes)

Naming convention standards enable teams to find configuration quickly and easily after it has been initially developed. Naming conventions of business rules, UI actions, and script includes should never contain information that is already accessible through other properties of the configuration. For example, the naming convention of business rules should never include capturing which table the business rule applies to, as this is a property that is already part of the business rule record and is searchable.

A naming standard that can serve as a starting point for any team is to simply document precisely what the configuration record is doing. For example, a UI action could be named *Set Resolution Action mandatory when State is Resolved*, which describes exactly what the client script is doing.

Script coding standards

Variable naming conventions, whitespace formatting conventions, where to put the curly braces when starting a loop, or an if/else block, there are numerous published coding conventions out there on the internet that can be adopted by the team when it comes to script coding standards. The most important thing is to adopt one as a team and attempt to follow it as closely as possible, taking care during code reviews to make sure any non-compliant code has been corrected. Coding standards and style guides improve the maintainability and readability of code on the platform, and the more teams and individuals developing on the platform, the more value it will have.

As there are numerous professional tech organizations that release their style guides for JavaScript (which are 90% applicable to scripting work in ServiceNow), in this book, we will not focus on our own style guide. We will take some time to highlight a few common elements across most style guides and leave your teams to adopt or establish the rest:

- Use clear, non-abbreviated variable names. Avoid variables such as `n`, `errcde`, and other aggressive abbreviations that will be difficult for people to interpret. Variable names should, as much as possible, clearly communicate their purpose: `errorMessage` or `accountId`. In most style guides, naming variables clearly and avoiding team-specific or group-specific abbreviations and acronyms are key naming considerations Saving typing time should be deprioritized over making code more easily understood by new readers.

- Minimize variable scopes. JavaScript contains many dangerous "features," with one of the more significant ones being the fact that variable declarations are at the global level or the function level. Most JavaScript style guides recommend that variables are declared as close as possible to where the variable is being used. With the Tokyo release of ServiceNow, the JavaScript engine will be able to use ES6+, which means local block-level variable declarations using `let` and `const` will become available. When your platform is on Tokyo or later, variable declarations should never use `var` to minimize the scope of variable declarations.

- Most popular JavaScript style guides use `lowerCamelCase` for variable and method declarations. Constants use `CONSTANT_CASE` and enumerations, and class names use `UpperCamelCase`. Private methods and private variables are a mixed bag. ServiceNow's own code tends to follow prefixing variables and method names with an underscore: `_privateVariable`. In comparison, other style guides prefer trailing underscores or no underscores at all (because JavaScript does not actually enforce any privacy; therefore, the underscore notation may, in fact, mislead developers into thinking that changes to these "private" methods and variables will not impact publicly available API consumers). This author's preference is to avoid the underscore notation considering JavaScript's limitations, but the convention should be set by the team and consistently followed.

Creating database indexes

Establishing some clear guidelines on when and how database indexes can be created can significantly improve the performance of your platform configuration, especially for high data volume scenarios in scoped applications.

Databases and database indexes are a science, but there are some general rules of thumb that technical teams can use to make sure that, at the very least, indexes are considered part of the technical design of applications.

First, indexing should be added to commonly used joins in queries against tables where the typical result set being returned is much smaller than the size of the table itself. Teams should strongly consider indexing whenever queries with joins return datasets much smaller than the table queried. In such cases, indexes will bring performance improvements even when the table is small.

Teams should frequently look at the **Slow Queries** log and then use the **Suggest Index** functionality of the platform to determine whether there are opportunities for improving performance through the creation of indexes for these queries.

Using scoped applications

Teams should be encouraged to use scoped applications for any functionality implementation instead of modifications to the global scope. Scoped applications have come a long way since their introduction, and as of writing this book, there are very rare situations where a scoped application would not be preferred over changes to the global scope.

Another aspect of scoped applications that should be contained within your platform development standards is how scoped applications should be structured. The most common approach is that each scoped application be a self-contained set of functionalities, configuration data, or data generated as part of the use of that functionality.

A single large custom application might still comprise multiple scoped applications, with each containing a standalone component of the scoped app. This pattern can be seen frequently with ServiceNow's own platform capabilities where each standalone capability that improves a core process (for example, the CAB workbench) is in its own scoped app.

One way to think about scoped applications is as microservices, with each scoped application providing specific functionality and a clear set of APIs and data tables designed for other scoped applications (or none) to interact with the functionality.

As with microservices, there is no single test for how much or how little functionality defines a "scoped application," but teams should at least try to design scoped applications in such a way that the vast majority (more than 90%) of functionality contained within the scoped app can be testable through the usage of simple mock data. This test will encourage teams to design strongly decoupled applications and simplify their scoped app's publicly facing APIs.

Putting it all together, we recommend that your platform teams commit to a standard where:

- Whenever possible, new functionality should always be created within a scoped application.

- Each scoped application should represent a clearly defined functionality that can be added or removed from the platform without interfering with other scoped applications.

- The vast majority of the functionality provided by a scoped application should be testable using mock data and stubs without requiring additional dependencies to be stood up to encourage good API, contract design, and decoupling.

Management of access controls and roles

When no specific design considerations exist, create at least one role per table that provides create, update, and read access to the table. ServiceNow creates a role by default when a new table is being created, and you should, at the very minimum, keep this *table-level* access role.

Table access roles should be rolled-up into *persona*-level roles (for example, Change Manager), which are roles associated with a specific user persona involved in multiple processes and/or user journeys) to control access.

Roles at the persona level should be assigned to groups, which users might be added to in order to be granted roles.

When adding roles to groups, always add persona-level roles and never table-level roles. The actual capabilities of the persona-level role can then be configured by adding child table-level roles to it. In this way, when creating a new scoped application, the team can define the access levels of the persona and provide the right table access for that persona by simply adding the appropriate table-level roles to the persona. This reduces the need for custom ACL rules or scripts against various user personas, as in most cases, providing personas with the right table-level roles will provide the correct level of access for the persona role.

Maintenance of design documents

A common refrain you hear from consultants and business leaders everywhere is that the documentation becomes obsolete the moment it is created. Too often this adage becomes a self-fulfilling prophecy as it is used to justify corner-cutting when it comes to the creation and maintenance of important documentation, which to the surprise of no one will result in obsolete documentation that.

The fact is documentation becomes obsolete if teams let the documentation become obsolete, often by choosing consciously or unconsciously to prioritize development throughput over maintenance activities.

However, *some* documentation can be critical to long-term value realization or achieving operational cost savings. The question is what documentation is important to keep and maintain and what documents should be optional or even avoided? Before we dive into which documents we recommend teams create and maintain, let us first go over a few guiding principles of documentation that inform our recommendations.

First, it is always good to create the least amount of documentation for the greatest number of use cases. This means that the documentation format and content standards are important – with the right standard and expectations setting, one or two document types can be created with care and then used to inform a myriad of use cases.

Second, whenever possible, have documentation be part of the natural workflow of implementation work and do not repeat documentation with the same purpose in multiple places. For example, establishing code comment standards and naming convention standards for configured platform components will serve naturally as documentation for the purpose of improving the speed of code understanding for new code readers. code readers with no additional external documentation. Similarly, enforcing the creation of user stories and acceptance criteria by the functional team to communicate required platform configuration changes to developers enables user stories to also serve as a paper trail of these changes on the platform, including when they are made.

Third, manage complex documentation with interdependencies to one another in a digital, searchable repository such as a wiki or a knowledge base. This allows dependencies between documents to be managed more easily. The most important capabilities to consider when it comes to managing documentation are the ability to keep track of (and revert/compare) versions of the documents, track who made any changes, and create easy-to-maintain links between documents. The Knowledge Management capability of ServiceNow could conveniently serve as such a repository.

Now that we have established our guiding principles, here are three pieces of documentation that serve 80–90% of an operational team's needs.

The **process design document** (or **system design document**) should serve as the bible of how the platform was designed to be used. This document should be in the form of a detailed process guide with every supported user and system action that is part of the instance design documented in a step-by-step manner. The process design document should explain the roles supported by the design, the processes and sub-processes supported by the design, and the platform interaction points and functional logic that is executed when users or automated actions perform these processes and sub-processes. Additionally, the design document should include details on how the platform and users executing the functional processes will need to handle errors. The process design document serves multiple purposes:

- The document can be used as a detailed reference for users of the system on how to perform actions and activities. When more documentation time is available and when end user experience is critical, this document's content can be summarized and formatted differently to produce training and end user documentation.

- The document can be used as a functional testing guide. Each process and sub-process captured in the process design must be tested via manual or automated test scripts, and the behavior described in the design document should serve as the expected result of any tests. During testing, if capabilities are identified that change the process as documented, it should be considered an enhancement, while issues identified that prevent the documented process from being executed correctly should be considered a defect.

- The documentation can serve to provide business context to development teams on required platform configurations. It is easy to reconcile this document with an agile methodology by treating each documented process/sub-process as the "Epic" user story that can then be subsequently broken down into actual user stories to assist in configuration.

The process design document may be maintained in lieu of the ongoing maintenance of functional requirement documents. This is because, after implementation, the process design document should meet all the accepted functional requirements of the project. This saves the team time and reduces the overall number of documents that need to be actively managed. Functional requirements that have not been met by the currently implemented platform should still be maintained or, at the very least, evaluated on a regular basis, but only as a function of **demand management**.

Architecture and **master data documentation** are needed and should be maintained with the purpose of helping technical teams quickly understand how pieces of functionality were implemented and their various components for the purpose of troubleshooting or making enhancements. Architecture documentation does not need to be overly complex and filled with diagrams. Instead, if process design documentation has been created, architecture documentation can be created that references processes and sub-processes and summarizes the components that enable those processes and sub-processes. For example, a process design document might contain a process for incident auto-resolution that also contains sub-processes for reopening auto-closed tickets. The corresponding section in the architecture document can then speak to the various platform components (a scheduled job that looks for stale tickets, the addition of a "reopened by user" metric, and more) that work together to facilitate the process. Architecture design documentation should be created prior to the detailed technical design of individual capabilities as it should serve as a guide for the technical design of individual components.

One type of content that must be documented in the architecture and master data documentation but is not obvious from the process design is the architectural design of shared components that are consumed by multiple processes. This includes any kind of foundational or master data and the data entered into the system following the **common service data model** (**CSDM**).

Test documentation should provide testing steps that testers can execute to test all, or parts of the design as captured in the process design document. Test documentation must be maintained in parallel with the design documentation and is used for regression testing on the instance. Testing documentation for manual testing should be maintained even as substantial portions of your functional configuration are tested via automation.

The preceding documentation forms what we consider the minimum required documentation to keep at the end of your platform implementation and during operations. Other documentation augments these core documents for specific purposes and, depending on the needs of your organization, the use cases supported by your platform. Some major missing but commonly understood documentation might require some justification. You might not agree with the logic, but you will at least understand the reasoning. It is not necessary to maintain user stories as their primary purpose is to serve as a communications tool in terms of what a developer should configure on the platform at the time of user story elaboration. While the user story documents themselves should be kept (hopefully in a digital repository such as ServiceNow), there is no need to maintain these user stories once the configuration has been completed and deployed into production. If enhancements or changes to the platform are required, update the corresponding design documentation and write new user stories instead.

As-built configuration documentation or detailed inventories of technical changes might be another documentation type that is commonly encountered, especially if your organization has engaged a consulting company for your initial implementation. While a detailed technical inventory of configuration changes is important for initial knowledge transfers to the operational teams, we believe it is less necessary to maintain this documentation in the long term provided that the platform team follows the conventions and standards of how to configure and name changes on the platform. For most simple features and configuration changes, the records in the ServiceNow instance itself and its descriptions and comments should speak for themselves. For more complex functionality with many moving parts, the higher level architectural and master design documentation should serve as enough of a guideline for technical teams to find what they need on the platform.

Automated test coverage standards and/or regressing testing scripts standards

As we covered in planning an implementation program for success, regression testing is an important part of any implementation. The bigger and more successful your platform team becomes, the more important regression testing becomes. For large implementations, regression testing existing functionality could cause significant bottlenecks to the pace of release if there is no automation available.

The automated test framework capability in ServiceNow allows the development team to create regression tests for business logic quickly and maintain them on the platform. The best way to ensure appropriate test coverage and make sure the automated tests are maintained is to require the creation or updating of automated tests at the time of the implementation of the configuration or change in configuration.

ServiceNow provides plenty of documentation on the uses and best practices of the automated test framework, so for this book, we will summarize a highlight reel of the most important best practices and throw in a few of our own words of wisdom to be applied to your testing standards:

- Use an automated test framework to perform the functional testing of business capability. That is, automated test frameworks are best used to test end-to-end user journeys, processes, and sub-processes and not the best for unit testing of specific script functions, individual business rules, and the like. In later chapters about how to operate the platform, we will discuss how you should keep the overall process steps enabled by your platform design documented clearly. If you follow that advice, it should be easy to create and group regression tests into the respective process and sub-process test suites and update them as those processes and sub-processes evolve.

- Tests should always be impersonating specific test users with specific roles; avoid writing tests that run with admin or non-real-world privileges.

- Set up and use the headless browser for UI testing. The headless browser allows the automated test framework to execute UI tests without forcing a manual opening of a browser window by the tester. Setting up the headless browser requires the installation of a Docker image on a server within your environment that can then host the headless browser for the ServiceNow instance, and the most up-to-date instructions can be found in the ServiceNow product documentation for your version of ServiceNow. Setting up the headless browser and making your UI regression tests utilize the headless browser is strongly recommended to enable UI regression tests to be executed with a minimum amount of human intervention.

- Treat test design as seriously as feature and functionality design in your implementation standards. It is *hard* to design good tests, and creating good functional tests requires careful thinking, planning, and design just like the creation of new capability. Whether the creation of automated tests is the responsibility of dedicated QA developers or the functionality development team, budget plenty of time for test creation and modification, with formal planning sessions and design documents created to make sure there's sufficient coverage of common journeys, failure scenarios, and error handling scenarios. When designing test cases, the team should be asked to map tests to functional requirements and/or the processes and sub-processes in your platform functional design document and use this mapping to determine the test coverage of the functional requirements. Strive for test coverage in the high 80%–90% range, making sure to cover the most common and important business processes first.

Now that the team has been set up with strong standards, in the next section, we will discuss the day-to-day operational norms of the platform team in the maintenance of your platforms.

Operational processes and techniques of technical development

What should your platform technical team's day-to-day activities look like? How do you manage your platform development team or teams? What should your team be watching for in their operational day-to-day and what should they be considering when they are doing development? In this section, we will go over a list of techniques for you and your team to keep in mind.

Teams should manage the accuracy of their estimations and the consistency of their throughput

Platform development teams in operations can become exceptionally accurate at being consistent with their estimates of effort and their actual throughput if the team puts their mind to doing so. This can apply to both teams attempting more waterfall approaches and agile approaches.

All that is required to estimate and determine a team's throughput is to have the team establish a unit for estimating average team effort (this could be story points, hours of average team effort, or some other metric that the entire team can agree to), a way of dividing work into chunks that can have a clearly measurable definition of done, and checkpoints where the number of work chunks that have been done since the last checkpoint can be measured.

A typical operational process that hits all of the preceding requirements is the scrum methodology, where user stories (with a clear definition of done and acceptance criteria) are used as a way of dividing work, story points the method of estimating effort, story poker as the way of obtaining the average team estimate and finally sprints as the checkpoint to calculate throughput.

We will not highlight the scrum methodology (or similar methodologies) in this book, but instead, we will highlight a few bad behaviors that can influence the quality and success of these estimated and throughput measurement processes and discuss how to manage them or use them to drive the continuous improvement of your processes. Far too often, scrum masters, project managers, and managers give up on attempting to measure the metrics or dismissing them outright when, in fact, there were clear external factors that they could influence to improve upon their ability to obtain good metrics.

One common bad behavior is an inability to recognize the impact of skill gaps between individuals on a team in the scrum process. The team and project management must recognize that the estimates for each chunk of work should be closer to the *median* estimate of the team, not the best or worst. In practice, this means that if the team member with the worst estimate is assigned the task, that team member should be supported (but not replaced) in the endeavor by someone (say, a team lead) to allow that team member time to grow into a better developer. One behavior to be avoided in managing this issue would be to have skilled resources do all the estimating, creating unrealistic estimates or estimates that leave no time for less experienced team members to grow and learn. A related bad behavior would be to allow less experienced developers on the team to repeatedly provide "infinity" estimates or announce that they cannot provide an estimate. If this behavior becomes prevalent, it will become not only impossible for the team to understand its throughput, but also prevent further investigations into why such situations exist. When the team or certain team members are providing many *I don't know* estimates, there could be several major reasons, each with different solutions.

First, it is possible that the requirement or user story provided is not detailed enough or is too large to estimate in an estimable chunk. These two issues are frequently related to each other – large features are unlikely to fit well into a single user story, and when a product owner attempts to do so, it produces unclear user stories. A good tell that this might be the cause is when the entire technical team has trouble with the estimation, and excellent product owners should recognize this as a sign that the story must be elaborated and broken down further to be estimated. As an aside, excellent product owners and technical team leads should make sure during the elaboration and estimation process that the "missing details" developers are looking for in this case are not technical in nature. Developers should be comfortable with taking clear functional requirements and translating them into a cohesive technical design hypothesis quickly. Developers should also be comfortable in asking for certain business contexts that then inform them of any technical choices that must be made without explicitly asking the product owner to make that choice for them. For example, when estimating a hypothetical functionality where *Sold Product* records that have expired must have a case created and assigned to the *Renewals* team, developers should be able to determine whether this can be done as a scheduled job occurring periodically scanning all sold products, or use an approach where each expiry event is pre-scheduled to trigger the task creation action as the *Sold Product* expiry dates are set. There is a possibility that a particularly complex or unique story might simply require some investigative time. In such cases, timeboxes should be established to prevent the team from getting lost in the investigation.

Second, it is possible that some or all developers might be inexperienced in the area in which the functionality has been written. If the estimates are consistently *unknown* for a set of user stories, the team simply might not have the skills required to tackle what is needed. In such a case, switching the functionality to a different team might be appropriate if available, and if not, the product owner and/or project manager should make sure that contingency in mitigating this risk is already incorporated into the project plan and budget.

Third, the issue could be one of skill or lack of confidence in that skill by a particular team member in that functionality. Good technical team leads should recognize when this is the case and decide on whether this could be an opportunity to upskill the team member or give the work to someone else on the team for now. In either case, during the estimation, the technical lead should, at the very least, provide some guidance on how a feature should be implemented to see whether the team member can, based on that information, provide a real estimate (even if high). Repeated *unknowns* by a single team member on multiple features over multiple sprints could suggest a major skills gap that should be addressed by the team lead. It is important that throwing out *unknowns* is not seen as something to hide, but instead as something that triggers proactive actions from the team to make additional investments to grow an individual's skill set at the expense of some team throughput.

Measurements of throughput also have common pitfalls, the most common of which is to give up after determining that the results returned are highly variable and difficult to use as a predictive tool. Instead, teams should look carefully into why throughput varies so greatly between sprint to sprint, or between two measurement cycles. Make no mistake – a team that has high variances in throughput from iteration to iteration should be strongly avoided as it means that there's little certainty on how much work will get done and no certainty in the team's ability to estimate. Teams should make it a clear agreed-upon goal to be consistent in their throughput and do everything they can to be consistent.

There are several reasons why throughput might be consistent. First, it is possible that the team's estimates are wildly off, and because throughput is measured by looking at the estimated effort, your throughput misses could just be your estimation misses. It is easy to determine whether this is the case: at each checkpoint, have the team provide the actual effort and determine the variance from the estimated effort. If this variance is found to be large, then the team might be too overconfident or not confident enough in their estimates depending on whether the variance is negative or positive. Either way, teams should be asked to improve their estimation confidence levels. There are various techniques available to allow teams to be better estimators. One way is by asking the team to visualize themselves being offered a bet where 90% of the time they win $5,000 and 10% of the time they win nothing, and then asking them whether they would take that bet over an alternate bet where they win $5,000 if the variance of the estimate is less than 10% (or 90% accurate). If there is a preferred bet, then the team is likely over or underconfident in their estimates as the odds should be equivalent. It turns out many studies have shown that mental exercises like this can improve the estimation abilities of individuals even without actually making the bet. Through this process of looking at variances and setting a team goal to reduce it sprint by sprint, behaviors will naturally arise within the team to encourage better estimates.

Another cause for throughput variance is distractions. The platform team might be called into meetings that have nothing to do with the work at hand, and be distracted by taps on the shoulders, incident investigations, and other activities. Distractions could be devastating to individual throughput as each interruption not only distracts the individual from the time it takes to resolve that distraction itself, but also the time needed to return to a mental state where the individual can be productively working on their development tasks. If estimations of individual units of work are right, then throughput variance is likely caused by disruptions and distractions. There might be some level of disruption that simply must be managed, but teams should not allow distractions from other work areas to significantly affect their ability to produce consistent throughput. If distractions due to other tasks are causing large throughput variances from sprint to sprint or cycle to cycle, scrum masters, project managers, and team leads must work hard to protect their team from such distractions and to help create behaviors within the team to avoid inter-team disruptions of throughput. Once again, by simply measuring and showing the variance of throughputs from cycle to cycle to the team and making the management of variance a goal of the team, many positive behaviors will naturally arise to reduce the variance and the cause of the variance. What teams must avoid is complacency and the acceptance of bad metrics. The moment a team gives up on controlling these KPIs is the moment improvement will stop or slow down substantially.

So far, we've talked about throughput *variance* as a metric, leading to the identification of process or skill gaps within the team. What, then, do project managers or team leads do about low throughput in general? The first element to consider is whether the concept of low throughput has any meaning. Generally, when project managers or team leaders feel that throughput is not high enough, they have in mind some theoretical throughput. If throughput variance has already been managed, this theoretical throughput is unlikely to be based on evidence from the team in question. Keep in mind that the metrics of throughput are only applicable to the team performing the estimates and the work; two isolated teams cannot be compared directly without more work of calibrating their estimates together as the true value of the unit of effort will be different across them under normal circumstances. So, when a team is concerned about its throughput being low, consider carefully whether the feeling is based on factual evidence or a dream of a theoretical throughput that the team cannot truly achieve. This is not to say teams cannot strive for greater throughput over time, but even this can be established as a goal of producing positive change in the team's throughput over time instead of a hard throughput metric goal that may or may not be realistic.

Teams should use version control whenever possible and use it as a way of managing code review and quality control processes

In *Chapter 6, Managing Multiple ServiceNow Instances*, we discussed in some detail how to manage your ServiceNow instances and environments. This instance management approach can be augmented with the use of version control.

In normal development not using ServiceNow, version control allows many people to develop and touch the same or inter-related functionality at the same time and distributes the work of conflict mitigation to each individual developer as they pull down the latest version of code incorporating the changes of other. In ServiceNow platform development, the usage of version control not only provides this capability but also has the added benefit of enabling individual developers to use personal development instances for real development with a much lower impact if the instance expires, as you would be storing your configuration and platform code in your own repository and simply using the instance as a development platform.

The simplest way to get into the usage of version control for your team is to use Git and GitHub as your repository of choice. This is because ServiceNow supplies native integrations through Studio to GitHub, making setup a matter of looking up the official ServiceNow documentation online and creating your repository.

When working with version control for ServiceNow, a suggested pattern to consider would be to set up a repository for each scoped application being developed. For the next few suggestions, we are assuming your team already has some experience or has learned some fundamental Git concepts. If not, go look for the vast amount of documentation online explaining how it works. Many of the suggestions apply just as well in non-ServiceNow development, and we encourage your team to look up best practices for version control outside of this book for added suggestions and inspiration.

First, teams should decide on a branching strategy or **workflow**. One of the most basic version control workflows can be called the *central repository* workflow. In this flow, everyone working on features will develop their features from the *main* code branch. Each developer merging their changes into the main code branch will need to incorporate features made by other developers to their own local repo prior to committing their changes to the tip of the main branch. Upon this basic structure, numerous branching strategies have emerged such as **Branch per feature**, where each feature is developed by developers against a unique *feature branch*, which, once completed, is merged into the *trunk*. This workflow provides the advantage of enabling releases to choose which features to include by merging individual feature branches into a release branch that is then committed to production. The following figure illustrates the branch-per-feature strategy:

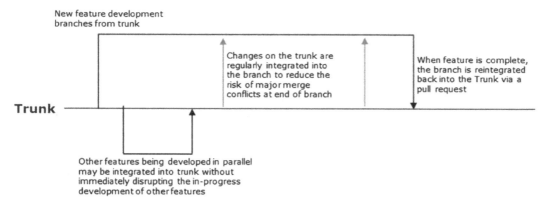

Figure 8.1 – Branching per feature

GitFlow, another popular branching strategy, involves the use of two major branches, one to keep an abridged version of final merged changes and another that is used as an integration branch for features. The advantage of this method is that the trunk branch contains a simple and clear list of changes, each of which is a minor or major version release, while the development branch can be used for commits and merges on a day-to-day basis and keeps track of a full history of changes to the complete code base.

The most practical advice that can be given with regard to branching strategy is to pick one that works for the team and be consistent in its application. There is no perfect branching strategy as each has its own drawbacks and advantages. Teams should take care and select the strategy based on the composition of the team and the types of work being done by the team. A small platform team making basic enhancements and defect fixes might use the *central repository* strategy as it is simple to keep track of and requires no further developer training beyond the basic usage of version control and its capabilities. In comparison, a large platform team where there is a team of developers per feature being built might benefit from the reduction of conflict resolution overhead by leveraging GitFlow or branch-per-feature strategies.

If you choose to have every developer use their own developer instances for the bulk of development, you should have a "Trunk" development instance that then serves as a snapshot of the *trunk* development branch – the branch that has a list of all completed development changes. This would include changes that are planned for, but not yet integrated with the QA or production environments.

Teams should make sure developers commit completed changes to their working branches (for example, the trunk in branch per feature or the develop branch in GitFlow). Committed changes will become quickly and inevitably integrated into someone else's local branches. This means that if you commit changes that are only partially complete, you will end up breaking other developers' local versions when they merge with your own changes. Therefore, it is important for developers to only commit to the main branch when your functionality is done and verified.

Summary

In this chapter, we covered multiple areas where it pays to establish operational standards, enforce, or highly suggest the use of technical and functional techniques to increase the effectiveness of your platform team's operations and deliver greater value more efficiently to the organization. As stated in the introduction, treat the contents of this chapter as a series of principles concepts for your team's toolbox. Some of the advice might not be directly applicable to your team, and others you might have already implemented, but whatever the case, we hope it has made you think and evaluate whether you are (or are not) doing certain behaviors and understand why they may or may not be done to see whether you need to make changes.

If there is a key takeaway to leave you with at the end of this chapter, it is that good technique in both the functional and technical world requires discipline to develop, and creating discipline is a matter of establishing cadences, measurements, and standards that can be enforced, and in setting meaningful and measurable goals. Before you implement any of the preceding techniques in this chapter or where you find them elsewhere, determine why they matter to you and your team. Make sure your team shares these goals and objectives, and you will be sure to succeed even if you run into obstacles and pitfalls at the beginning.

Setting your platform standards will help your platform team produce high-quality, maintainable configurations. In the next chapter, we will discuss how to structure and engage the rest of the organization to maximize the return on investment of your platform.

Part 3 – From Success to Innovation

Building on everything we've learned so far, the final section of this book will address the innovative possibilities that can be unlocked on a ServiceNow journey and help you navigate the uncertainty and claim the value of being an early adopter.

This part of the book comprises the following chapters:

- *Chapter 9, Effectively Operating ServiceNow*
- *Chapter 10, Artificial Intelligence in ServiceNow*
- *Chapter 11, Designing Exceptional Experiences*

9
Effectively Operating ServiceNow

So far, you have deployed ServiceNow in your organization, and if you have followed the advice of the previous chapters, it went exceptionally well. The scope was well controlled, the stakeholders were supportive and aware of the decisions they needed to make, and the overall direction of the implementation was smoothly steered by a clear overall strategy and guiding principles: congratulations!

At this point, the project is over, the consultants (if any) have left, and the current users of the platform are starting to ask for changes and reporting little issues they are finding here and there as they become increasingly familiar with their new world. In addition, leaders from other departments have heard of the success of the implementation and are clamoring to start using the platform for their own needs. One executive in marketing heard that there's a Customer Service Management module on the platform and they're eager to leverage the platform to send targeted marketing campaigns to customers, augmented with case data. With half or more of the team involved in the initial project gone or reassigned to other tasks, the demand is piling up and frustration is building: what do you do?

This chapter will help your organization *operationalize* the platform for long-term success and value realization far beyond the initial implementation. The following major topics must be considered to make operationalization successful:

- Establishing a platform strategy
- Establishing a platform operating model
- Operating the platform

Before we discuss these important dimensions of operationalization, let's briefly look at the difference between *operating* the platform and simply implementing capabilities against it.

The differences between operations and project work

While platform operations and project work can intersect in many areas, there are several key differences between the two that require different models to be managed correctly. The following is a list of the key activities for both project and operational work and highlights their differences:

- **Business outcomes**: Both project work and operational work delivers specific business outcomes and, in this respect, there is very little difference between the two. Platform operations will more frequently involve "ongoing" business activities (defect remediation, incident management, platform performance, and maintenance management) that occur continuously and are typically not always accounted for during projects.

  The biggest difference between delivering the business outcomes of operations and projects is that operations are typically **resource-constrained** while project work tends to be **time-constrained**. That is, projects will allocate as many resources (money and people) as necessary to deliver an outcome based on a specific timeline, whereas operational platform work generally has a fixed resource capacity but can work at delivering a business outcome continuously until it is achieved. It does so by increasing timelines and staging value realization across multiple regular releases.

- **Timelines**: Project work typically has fixed timelines to deliver a specific outcome, whereas operational work must be more flexible with timelines because it is resource-constrained. The advantage of project work is that it can deliver major changes very quickly by concentrating its resources on a specific scope. The advantage of operations is that they can work on a specific business problem for as long as they need to, which can significantly improve adoption and can be more successful for changes with many dependencies.

- **Resources**: Projects and operations both tend to have fixed budgets. However, projects are generally funded based on a business case meaning that an additional budget is allocated to achieving the targeted Return on Investment. On the other hand, operations typically have a fixed annual budget, a portion of which will be dedicated to ongoing activities such as incident management, platform maintenance, and defect remediation. The remainder is flexibly allocated to enhance the platform.

Because of these differences, platform operations is much more sensitive to how demands are prioritized than projects – they must understand how to allocate its resources across a portfolio of possible demands so that they can still deliver business outcomes regularly, avoiding situations where they spread their resources too thin. Operations must also worry about how the ongoing costs of maintaining the platform can be affected by their activities, as well as those of the projects. Depending on the funding model for platform operations, all these ongoing costs must be absorbed through the operational budget, so platform operations is also far more sensitive to strong rules and standards when it comes to platform governance, without which they may find themselves overwhelmed by projects whose outputs must be maintained in perpetuity.

Operations need a "North Star" to guide the way it prioritizes and allocates its resources; one such North Star is the platform strategy.

Establishing a platform strategy

The first step in determining how your organization's ServiceNow platform will be effectively operationalized is to determine what your platform strategy will be. ServiceNow, when used at its fullest extent, can be an enterprise platform that acts as the source of truth for several types of operational data in different process areas, a platform that supports multiple applications used by different teams, and a significant (if not the only) point of entry to engage business services and customer services. With so many capabilities, having a platform strategy will help focus your efforts so that the organization's limited resources can be directed against the platform in focused ways that drive business value.

But what even is a platform strategy? You will often see a platform strategy or technology strategy being summarized in a single statement, such as *Automate customer service* or *Decrease the cost of IT*. While these strategic statements can be useful to help simplify and specify the goals of your team or organization, having an effective strategy that can help you determine how to design the operating model of your platform will take a little more work.

Instead of racking your brain to find one magic statement that perfectly describes the objectives of your platform, you should consider and provide answers for various strategic dimensions. Let's take a look at some of the questions that are typically asked.

Who will be the platform's customers?

When you launch the ServiceNow platform into the organization, it may seem obvious who the platform's key customers will be. Typically, it is the team(s) of the executive sponsor of the implementation or their business customer. However, once the platform has transitioned into operations, the line can rapidly blur as groups related to the platform's implemented capabilities begin requesting enhancements and changes to the platform to improve its organizational value. This influx of demand from various stakeholders is a good thing: it typically means that the implementation of the capability has been successful enough to garner attention and there are now opportunities to improve the return on investment. However, if the platform strategy does not identify its priority customers, the platform operational team's resources may be stretched thin, and the demand intake process may be overwhelmed.

By identifying an initial list of priority customers and growing that list over time as the platform expands (or not), the platform team can manage expectations and better focus its resources on delivering the right outcomes. Given the capabilities of the ServiceNow platform, there are several immediate customer candidates and use cases the platform can support:

- **Chief Information Officer** (CIO) and IT service management leader
- Global service desk leader
- **Chief Information Security Officer** (CISO)

- Head of HR

- Head of customer support and customer service

- **Chief Operating Officer (COO)**

Each of these leaders and titles is likely to oversee a team(s), departments, or a portfolio of business use cases that can be directly addressed by the implementation or enhancement of the platform. More importantly, the portfolio of business cases that exist within the teams of each leader in the preceding list is likely to be large enough to justify long-term, ongoing investments on the platform. If your immediate interested customers are more operational – for example, the incident management leader or change management leader – you should work with them to build a broader business case for the platform by including their leaders or broader stakeholders. As a platform with broad enterprise capabilities and a fairly substantial operational cost, it is important to help stakeholders at the operational levels to build broader business cases to make both them and the organization successful.

Identifying your customers will not only focus the efforts of the platform team but also make it easy to measure and maximize another important metric in the long-term operational success of an enterprise platform: customer satisfaction. Without a clear customer and long-term sponsor, the platform can quickly run the risk of failing to make anyone happy. When such a situation occurs, leaders and teams can quickly decide to switch to something else, washing away all the time and effort invested in the existing technology.

The platform's customer base can be added to over time, so it is always good to start with a smaller customer base before expanding it. A good rule to follow here is to treat the operations of the platform as a product business within the business. Like most successful businesses, focusing on doing one or two things well at the beginning is typically better than tackling an array of different and distinct issues. Also, like most businesses, there is still an element of opportunism that must be considered. When a major customer with a large and high-value use case approaches, there may be a strong incentive to consider that customer, even if it means a pivot in the direction and focus of the platform team.

When considering this question, the answer will usually come down to the following:

- First, who are the current customers that are utilizing the platform? In most cases, these customers must be directly supported and therefore are likely to be your platform's core customer base for the reasonable future.

- Second, is there a broader strategic initiative that the platform can play a key role in supporting? For example, if a large customer service transformation initiative with ambitious goals is currently being planned, the ultimate stakeholders of that transformation could easily be considered potential customers for the platform.

- Third, who are the prospective near and medium-term customers who may need to be engaged early and whose relationships should be managed for long-term platform success? To identify who these customers are (and what level of involvement they should have in the ongoing operations of the platform before being a customer), the second aspect of platform strategy should be considered: how ServiceNow fits within the business.

Even if you have understood your customer base, it may not be clear where ServiceNow fits within the broader technology landscape of the organization. In the next section, we will see how positioning ServiceNow in the broader technology landscape is important to successfully plan the platform's growth.

How will the platform fit within the broader technology landscape?

Whether it is IT or HR or the customer support and marketing parts of the business, you are likely to find a broad array of technologies and platforms available to solve a large and broad portfolio of business problems. Some of these technologies will be fit-for-purpose solutions built to be exceptionally good at solving a particular business problem, while others, such as ServiceNow, can act as an enterprise rapid application development platform capable of being configured to address multiple business cases. To make this situation even more complex, ServiceNow also has out-of-the-box applications and capabilities that are fit for purpose, so even if the rapid application development aspect of the platform is not leveraged, ServiceNow can still be considered when solving business problems in the HR, customer service, IT service management, and other spaces.

The question to answer via your platform strategy is how the enterprise should treat ServiceNow. Is it simply another tool in the toolbox, or will it be treated as a more strategic asset, taking advantage of the fact that, as a platform, return on investment can often be greater as more use cases – in particular, interconnected use cases – are addressed on that single platform?

When treated like a toolbox, for any business scenario that could benefit from supporting technologies, ServiceNow will be evaluated against the rest of the technology portfolio of the organization and potentially against off-the-shelf and non-off-the-shelf solutions not yet within the technology landscape of the organization. This is a good choice if the organizational technology strategy is to emphasize fit-for-purpose solutions for each business outcome. The disadvantage of this strategy is that the cost of integration typically increases, and the approach may introduce a multitude of point solutions that can increase the overall technology costs. The resiliency of the technology landscape may also begin to erode, especially if introducing many different solutions overwhelms the capability of the organization to retain the specialized skill sets that may be necessary to keep those technologies working effectively.

When treated as a strategic platform, when a business use case is identified, ServiceNow is treated as a priority solution option (along with a small selection of other strategic platforms). In this way, the question that's asked when a new business case is identified is *why couldn't this be solved with ServiceNow?* as opposed to investing heavily in identifying the potential *best* fit-for-purpose solution that may be out there. The advantage of this strategy is that it maximizes the potential of the *platform* to share costs and to deliver integrated value more efficiently. When supported with the right platform solution architecture, leveraging ServiceNow in this way can significantly reduce the cost of integrations and decrease time-to-value, enabling the organization to launch applications that can take advantage of an enormous amount of enterprise data with less effort. In addition, leveraging a small set of powerful enterprise platforms allows the organization to maintain a smaller core of skilled talent that can be flexibly moved across different applications and technology solutions built on the same platforms The disadvantage of this approach includes potentially losing out on some functionality that is more readily available on fit-for-purpose platforms and, in some cases, a reduction in flexibility for business stakeholders looking for quick solutions for business problems that may be more readily available and more readily deployable when choosing an off-the-shelf point solution.

While both strategies are valid, ServiceNow's capabilities, costs, and long-term roadmap strongly encourage organizations to treat it as a strategic platform. While ServiceNow may be competitive from a value standpoint against point solutions, a far more compelling value case is made by treating it as a strategic platform.

With that said, there may be internal limitations within the organization that prevent the platform from realizing its potential, even if there are pockets of sponsorship. These limitations should be considered when a strategy is being established so that the strategy is realistic for the organization. In the next section, we will discuss one of the most impactful determinants of how successful a particular platform strategy will be: the platform operating model.

Establishing a platform operating model

The best and most ambitious strategies will still fail miserably if the enterprise is not organized or resourced enough to successfully execute that strategy. Decisions on how the ServiceNow platform is governed can often affect the ultimate direction it takes during operations, regardless of the strategy it is attempting to follow. Therefore, it is important to consider the operating model when the platform strategy is being developed. Leaders must recognize cases where the ideal operating model is not supported and adapt accordingly. Given existing organizational realities, it may be prudent to shift the strategy accordingly to achieve more realistic goals. On the other hand, it may be that success in achieving a particular strategic goal is imperative for the organization, so leaders will need to put in the work to build support and governance to execute on that strategy.

While there is no one-size-fits-all approach, like strategy, there are a few key aspects of the operating model that should be decided on to create a coherent approach:

- Who will manage the direction of the platform and how?
- Who will manage the intake of platform demands and how?
- Who will manage how the changes on the platform are implemented and how?
- Who will manage the funding for changes to the platform and how?

For each dimension, there is usually a *centralized* versus *distributed* option with pros and cons for each.

Before we look at each of these models and how they differ, let's look at the typical roles involved in platform operations, from setting the strategic direction of the platform to executing the changes required to meet business outcomes.

Platform operating model roles

Whether big or small, centralized or distributed, each platform team should fill a set of roles that can help the team operate effectively. A person can have multiple roles in a team, but the team should avoid splitting the responsibilities and accountabilities of a single role across different individuals. This is because the accountabilities and responsibilities separate concerns in such a way that individuals in a role can fully focus their abilities on related areas and create appropriate separation of concerns that drive better decision-making as a whole.

Platform executive sponsor(s)

The platform executive sponsor could be anyone (or several individuals) within the organization. The primary role of the platform executive sponsor is to fund the costs of platform operations and, occasionally, the costs of projects on the platform as needed. Because platform executive sponsor(s) fund the platform operations, they have a considerable influence on the direction of the platform and how operational resources are allocated. However, platform executive sponsors are not omniscient. Since they are leaders within various business areas of the organization, they may have competing priorities and interests. They are also likely to only have a limited understanding of the operational constraints faced by the platform team. These constraints may include resource constraints as well as technological constraints and dependencies on other parts of the organization. Because of this limitation, another role must exist with a greater level of understanding of these constraints to provide insights into these areas and improve the ability of executive sponsors to make decisions. This role typically belongs to the platform owner.

ServiceNow platform owner

The ServiceNow platform owner is an individual that oversees the platform operational team and is accountable for the health and business outcome realization of the ServiceNow platform. The platform owner should be well informed of the operational details of the platform – how much work will it take to maintain the platform and what kind of budget is required? What are the current risks? The platform owner must be able to advise the executive sponsors on how to maximize the value of their investment, providing a level of understanding of the operational realities that the sponsors may lack. There should *always* be a ServiceNow platform owner, regardless of whether the platform is governed in a more distributed manner or a centralized manner. The only difference in such a case would be the authority provided to the platform owner, as we will discuss in more detail in the following sections of this chapter.

Business analyst/product owner

The names business analyst and product owner are often interchangeable; organizations following agile terminology tend to use product owners more than business analysts. The actual job of the role is the same – ideas and needs from the business that may be unstructured and ad hoc must be collected and refined into a clear and formal request with well-defined business outcomes and requirements. The business analyst or product owner's role is to understand the business needs of the customers and translate their ideas and unstructured demands into structured demands/requirements that can then be scoped, estimated, and prioritized as work to be done by operations. The best product owners understand the customer's needs better than the customers themselves, make prioritization decisions that are best for value-realization for the customer, refine their ideas into clearly executable demands, and represent the customer in the demand management process. When a demand is approved and prioritized, the product owners flesh out their requirements into target state designs and user stories, which they then work with alongside the platform developers to estimate and ultimately implement them.

Demand managers

Demands from the business, represented by their respective product owners, must be evaluated and qualified against other demands and operational constraints before they can be officially scheduled as a package of work for the operational team. Demand managers are the first to enforce platform governance – they enforce the standards that the demands on the platform must meet before they can be formally assessed and prioritized. Demand managers will interact with the product owner to determine if the demand is sufficiently thought out so that it can be evaluated against other demands. They can also reject demands that do not align with the strategic priorities of the platform outright. Depending on the volume of demand and the footprint of the platform, there may be many demand managers managing demands against the platform. Demand managers are responsible for the platform owner in a centralized operating model as they help the platform owner determine how best to allocate operational resources to maximize business outcomes.

Platform architect

The platform architect is responsible for protecting the architectural integrity of the platform and provides the platform owner with technical insights that may affect the prioritization of demands against the platform. A common role of the platform architect is to evaluate and reject demands that may not be the right fit for the technical capabilities of the platform (for example, requirements for the platform to be a data archiving solution for enterprise documents or a training video viewing platform). The platform architect also provides demand managers and the platform owner with support in assessing the relative complexity and/or technical risk of implementing a demand and the long-term operational costs and risks associated with maintaining a particular demand going forward. Platform architects also play a role in project work, using the same framework in which they assess the risk of operational demands to projects, as has been detailed in previous chapters of this book.

Platform development and QA teams

ServiceNow developers, QA testers, and their respective leads comprise the platform implementation teams. They are responsible for planning and executing the necessary configurations and changes on the platform to meet the incoming platform demands. They are also responsible for upgrading the platform, responding to platform incidents, and monitoring platform health. The size and number of these implementation teams vary, depending on the platform's footprint, but at a minimum, there must be a technical lead and a QA lead on each team. One is responsible for implementing the technical configuration necessary to meet the requirements for approved demands, while the other is responsible for testing these changes against the acceptance criteria to ensure the output delivers upon these business outcomes with minimal defects. In large operational teams, each lead may have a team consisting of two to four individuals to distribute their work, and/or there may be multiple such teams each responsible for its stream of work. This is typically divided by the product owner/business line they work with.

In the case of large implementation teams, the role of the technical lead is to provide technical guidance and direction to their team so that every developer implements them according to the right technical standards. The technical lead also works with the architect to ensure that what is implemented follows architectural standards and considers the usage of the shared architecture (for example, CMDB) in a way that is common across multiple streams.

The QA lead is responsible for understanding what configuration is being built and working with their team to construct the right test scenarios to fully exercise the configuration. In large and complex operating environments, the QA lead and their team are also responsible for creating automated tests and other testing technologies to enable the operational team to perform regression testing and automated testing for complex configurations.

Release managers

Release managers help plan when capabilities will be deployed onto the platform. This job is complicated by the fact that there may be external dependencies outside of the platform that provide constraints on when releases can occur (for example, there could be an organizational policy that dictates no changes to the HR module during the year-end payroll processing period). Therefore, release managers work with the broader enterprise release process to determine the best sequencing and timing for releases of capability and help demand managers, product owners, and development and QA teams sequence and schedule accordingly. Release managers are also responsible for helping create the deployment plans for operational activities that are ready to be deployed by working with the development and QA teams and the platform architect.

Regardless of the exact operating model, these roles will always be needed to manage platform operations. But while the existence of these roles is a requirement for healthy platform operations, the exact authority and responsibilities of these roles may be subject to change, depending on the key dimensions of the platform operating model, starting with how the direction of the platform will be managed and by whom.

Who will manage the direction of the platform and how?

Given limited resources but many demands on the platform, who should decide which demand is answered first? The answer to this question impacts the authority that's granted to the platform owner and may also impact the speed with which platform decisions are made.

In the *centralized* model, the platform owner has the ultimate authority to decide how resources at their disposal will be allocated against demands.

When the right individual is chosen and provided with the right team, the centralized model tends to make faster decisions A centralized model where the platform owner has the final call on the direction and vision of the platform can also be more effective at solving a large problem if provided with enough space, as the platform owner can focus their resources on realizing a particular vision without being distracted by the changing priorities of many different stakeholders. The drawback of the centralized model is that it is more difficult to find the right platform owner as the skill requirements of the role are increased. The platform owner must balance selling the platform's vision to stakeholders to receive funding with managing the resources of the operational team and managing and selling the business outcomes. Therefore, the platform owner should be someone senior within the organization who is deft at building consensus as well as building a team that can deliver promised results.

In the *distributed* model, the direction of the platform is decided by an **operating committee** that should meet regularly (the cadence of which depends on the anticipated and actual volume of platform demand) and decide, as a collective, which demands should be implemented on the platform first, which demands should be deferred, and which demands should be rejected outright.

Under the distributed model, the platform owner acts as an advisor and steward of the resources and direction of the platform granted by the operating committee of the platform. They are just one vote among many regarding how resources should be allocated, and which demands should be addressed first and which second. When the stakeholder of the operating committee is aligned on following a particular shared vision and acting in unison for the interests of the enterprise, the distributed model has a better chance of finding greater transformational opportunities by pooling resources from multiple leaders for the greater good. The drawback of such a model is that without an organizational culture of working for a common good, or without the ability to measure the outcomes of that common good, it can be prone to paralysis as competing priorities and objectives fragment the decision-makers, which ultimately results in inaction.

The platform owner in a distributed model should still be as senior as possible, as their capacity to act as an advisor to the operating committee is enhanced by their seniority. However, because the platform owner is not ultimately accountable for the direction of the platform and instead acts as a steward or executor of the will of the operating committee, the platform owner could be a less senior resource in the distributed model compared to the centralized model. In such cases, it is recommended that the platform owner brings a stronger execution capability to the table to bolster the more strategic capabilities of the other stakeholders within the operating committee.

In either model, the platform owner is responsible for managing the expectations of the stakeholders and the operating committee.

While there is no wrong choice in terms of whether your platform operating model should be more distributed or centralized under the platform owner, there are factors that should influence the decision. If the organization is highly effective at measuring enterprise metrics and can incentivize leaders to deliver enterprise outcomes, even if it's outside of their immediate portfolio (for example, incentivizing IT leaders to produce HR business outcomes), a distributed model may work very effectively. On the other hand, if the organization tends toward more of a siloed approach, where sharing resources and funding is not formally recognized, then having centralized accountability of the product owner enables better business outcomes by allowing the platform owner to make the right decisions on behalf of the combined inputs of the stakeholders as a "neutral" third party.

Establishing the operating model to determine the direction of the platform helps establish the framework for which prioritization decisions are made. The next important dimension to manage is how the source of these decisions – demands – is managed.

Who will manage the intake of platform demands and how?

Demands are requests to commit resources to make changes on the platform. Defects on existing capabilities that should be remediated, enhancements on existing capabilities, and net new capabilities to be implemented on the platform all constitute demands upon it.

Demands can come from anyone and anywhere, and as the footprint of the platform grows, demands for change typically increase. Apart from this, stakeholders become invested in how the platform works and is changed. Because resources are finite, only a subset of demands can be accommodated for any fixed period, so managing what demands should be accepted and which should be deferred or rejected becomes a key aspect of the platform operating model. Since we have already discussed establishing how the direction of the platform will be managed, the *who* for which demands are prioritized or rejected has already been established – the platform owner and the operating committee should ultimately make such decisions based on the direction they have established. However, this still leaves us with the opportunity to determine how to filter demands so that only highly relevant demands are fully evaluated. This demand management activity is required as without it, demands big and small must all be evaluated at the highest authority level, creating a bottleneck, and potentially overwhelming the decision-making authority. This results in less effective and informed decision-making.

It is convenient to look at how demands should be filtered and qualified by separating them into a few buckets – demands that enhance or remediate existing platform capabilities that are already being delivered to the organization, demands that are net-new but still aligned with the platform direction established by the platform authority, and demands that are net-new and add to the scope and mandate of the platform.

Demands that enhance or remediate existing capabilities should almost always be given some greater priority when business outcomes and costs are equal. This is because these demands already have clear customers and clear benefits and are already well aligned with the platform strategy and the broader business strategy (assuming everything leading up to this chapter has been followed!). In the case of these demands, the hardest problem is to size the demands and determine how to prioritize them if the overall number of demands is greater than the ability of operations to deliver them.

Demands that are net-new but still aligned with the direction established by the platform authority typically require an additional consideration – the increased operational commitment to support the capability. These types of demands include "major" enhancements that add substantial new capabilities to something already implemented – for example, adding a call center integration component to an existing customer service management capability or adding a virtual agent to the employee service portal. In the latter case, even though the capability can be added "out-of-the-box," adding it must still be assessed as it will create a source of additional demand, should employees or the business stakeholders begin noticing gaps in the service experience and begin asking for improved conversational options and other additions through the demand channel. A greater level of scrutiny is required when these new capabilities are added because additional operational strain may be placed upon the platform team, and because such additions may also have dependencies outside of the platform (for example to integrate with new data sources within the organization). A greater level (and duration) of funding commitment may also be required before they can be approved.

Finally, demands that are net-new and are not in the immediate direction or scope of the platform require the greatest amount of scrutiny and may also require one or more projects to be fully implemented. The biggest decision for these types of demands is whether expanding the platform's footprint into that area is worth it. Leveraging the platform and business strategy as a guide is important in these decisions – supporting a net-new business customer with a brand-new platform capability can create a large change in the operational constraints of the team and should always be carefully considered, even when the initial implementation can be projectized instead of being dealt with operationally.

Once a small subset of demands has been qualified, it must be considered against organizational and operational constraints before it is approved for implementation. During this decision-making process, *how* the operating model deals with resourcing for implementation can influence the evaluation criteria.

Who will manage how changes are implemented on the platform and how?

While we clearly articulated the roles needed to operate the platform earlier in this chapter, we have not specified the organizational hierarchy in which these roles fit. This was done intentionally because different organizations may benefit from different reporting models, and different decisions regarding whether the platform is managed in a distributed or centralized manner can also affect the reporting structure of the various roles.

Two major patterns are used in the organizational hierarchy of the operational team of the platform when it comes to using resources to implement changes on the platform: distributing the implementation team across the business stakeholders or centralizing the platform implementation team under the platform owner.

In the distributed pattern of implementation, business stakeholders are allowed to resource for their implementation teams. This can often be seen in organizations where business units such as human resources or security or customer support have their own technical teams to maintain systems that underpin the business services in their portfolios separate from IT. This pattern can also sometimes emerge when IT has subunits with technical delivery capabilities. These subgroups can often leverage their technical capabilities to bring enhancements to the platform that may otherwise be prioritized lower by the platform owner, creating parallel streams of value and reducing time to value for those teams. The advantage of the distributed implementation pattern tends to be time-to-value: groups that can find implementation resources can quickly meet their demands on the platform without worrying about broader operational constraints or impacts.

The distributed pattern has several drawbacks. As an enterprise platform, the value realization of ServiceNow improves when various components of the platform are configured to work together or utilize shared data and architecture. While this is possible to achieve when implementations are spread among multiple groups, it is certainly much more difficult. Even when strong governance and standards are established, due to the lack of direct reporting structures, it is extremely easy for that governance to be violated or ignored both intentionally and unintentionally. Therefore, the benefits of the parallelization of platform change work are hindered or mitigated by greater risk to platform integrity or a reduction in efficiency due to the overhead of attempting to coordinate these different teams into a cohesive whole. Related to the previous point, there are also disadvantages and inefficiencies in resource sharing as each group in a distributed model is usually more strongly incentivized to make sure investment and resources are allocated to benefit the group directly. This difference in incentives means that the distributed model tends to be less effective at pooling investments that may only have short-term benefits that are limited to a particular group, but a longer-term benefit for multiple groups. The CMDB is a notable example of this – initial investment in building up a high-quality, well-maintained CMDB may have many phases where only a few groups benefit, but over the long term, that investment can translate into greater value realization for all.

The alternative to the distributed approach is centralizing the implementation capabilities under a single leader, most typically the ServiceNow platform owner (and by extension, the operating committee if the platform direction is managed in a distributed way). In this model, the implementation team reports directly to the platform owner, and it is the responsibility of the platform owner to manage the performance of such a team. The centralized model truly makes the business stakeholders customers and removes them from having a direct impact on the platform, regardless of their ability to bring human resources to the table. The advantage of this model is that the direction of the platform can be much more easily connected to the priorities of the implementation. As the implementation team reports directly to the platform owner, who is a key stakeholder in setting the direction of the platform, there is little room for the implementation team to deviate from the long-term platform vision and plan. Additionally, the fact that the implementation team directly reports to a single authority means that it is much easier to enforce technical and functional standards on the team and hold the team accountable for the quality of their overall delivery. The centralized model is particularly good for efficiently and effectively delivering large-scale transformations against shared business components such as the CMDB. Assuming the directional alignment exists (a difficult step in and of itself), the implementation team can plan across its entire footprint to maximize the usage of the shared data and architecture and coordinate the timing of releases to take advantage of the CMDB's growth at the right times.

The disadvantage of the centralized approach is that there is a greater demand from the platform owner or the platform operating committee on managing its business customers and stakeholders. Because the business customers will continue to have competing priorities and when these priorities are addressed (or not at all) entirely at the whim of the platform owner and/or operating committee, at least some stakeholders may find their priorities de-emphasized for those of others. This disconnect will place more pressure on the platform owner and the operating committee to do much more in ensuring stakeholder alignment and that everyone understands and buys into the shared vision. It's also easier for the centralized approach to get lost in its vision and become disconnected from the customers they are trying to serve; this issue can be mitigated by having strong product owners who can act as the voice of the customer and help place the right focus in the right areas.

Concerning how to set the direction of the platform, there is no obvious choice between the distributed and centralized models to manage the implementation of changes on the platform. With the implementation choice, if all other factors are not considered, the centralized model tends to be a better fit for ServiceNow, given its nature as a shared enterprise platform. However, there are other factors to consider, and having a strongly centralized model will require several key ingredients, such as a strong platform owner and a highly aligned and supportive operating committee that can help the platform owner gain alignment with the business customers. This does not mean organizations should give up on the centralized model. Even if some distribution of responsibilities must occur, strong enforcement of architectural and quality standards through the platform owner and operating committee is still necessary for long-term value realization. The bottom line is that when it comes to implementation on the platform, leaders and teams must learn to work together in developing shared visions of both business outcomes as well as architectural approaches and keep each other and their teams aligned to this shared vision. This cohesion and clarity of where the platform must go from all stakeholders becomes invaluable when it comes to the next consideration for operating the platform: funding.

Who will manage the funding of changes on the platform and how?

Who pays for the costs of platform operations and what considerations must be taken to fund the platform? We can break down platform operational funding into three major buckets – the licensing costs of capabilities of the platform, the maintenance and enhancements costs of capabilities already implemented, and the initial implementation costs of the capabilities.

In many organizations, the business approaches IT with a particular problem or requirement, and IT is responsible for obtaining the necessary budget and resources to accomplish this task. IT in this model may also be responsible for all ongoing operational costs and the cost of defect remediation and enhancements going forward for the capabilities that have been implemented. This *IT Service Provider* model gives IT, which is the part of the organization most well-equipped to implement and operationalize the platform, great flexibility to prioritize and execute the initiatives that it believes can make the greatest business impact, considering all its potential customers and business opportunities. In this model, IT can charge its customers for ongoing operational costs, adjust that cost depending on the level of service required, and request additional funding for large projectized outcomes, provided the customers are willing to pay for it. Any excess earnings through this charge-back model can then be reinvested into IT for long-running, broadly impactful improvement initiatives such as improving automation or adding resiliency.

The disadvantage of the IT Service Provider model is that unless there is an actual mechanism for IT to charge its customers for its services, operationally and otherwise, there may be a level of disconnect between the budget and resources that IT can obtain, the business outcomes desired by the customer, and the actual outcomes that IT can deliver with its budget. This is especially true when it comes to long-term operational costs, which are often neglected and result in decreases in service quality and an inability for IT to reinvest back into itself regularly for continuous improvement.

The alternate model, where businesses must provide funding to IT to accomplish its goals (IT as an implementation team for the business), can make it easier for IT to execute focused initiatives. The model also encourages the issues with the most pressing business priorities (that is, the ones that people are willing to pay for) to be addressed first. When the platform team/operating committee establishes strong standards for what funding dimensions must be considered (for example, not neglecting the cost of maintenance and the cost of ongoing IT service management for the new service), the model may create greater delivery capacity than in the service provider model because every initiative will be backed by both the business and IT.

One of the major drawbacks of this model is similar to the issue with distributing the implementation to the stakeholders – the inability to commit shared resources to build capability on the platform that may provide long-term benefits for multiple areas. While strong governance and standards can reduce the instances of this occurring, it can impact the customer experience – business units with money may not understand (or even care to understand) the reasons why their goals cannot be accomplished without additional investments in shared infrastructure for which their funding is insufficient, creating misunderstandings or a perception that the platform is unable to deliver results.

The reality is that it may be difficult for the ServiceNow platform stakeholders to make any tangible changes to the funding model as this is typically a deeply embedded aspect of business operations. Now that we know the players of the game and what to look for in terms of platform operations, we will focus on the day-to-day actions of the team and where to apply their focus, including how and where to focus the time of the platform team (including the operating committee) to reduce the impacts of the drawbacks of each model and to play to their strengths.

Operating the platform

Now, we are going to highlight some of the key activities and actions members of the platform operations team can and should be doing to contribute to the platform's success. This will not be a detailed set of operational procedures, but instead a loose list of actions and considerations for each role to be utilized as a guide to drive everyday positive behaviors. Where applicable, we will provide comments on how a particular activity may change in terms of its criticality or character, depending on the operating model or choices established in the previous sections of this chapter.

Executive sponsor and operating committee

The executive sponsor and the steering committee have three major responsibilities that they must fulfill.

Protecting and committing to the platform's vision and strategy

One of the most important aspects of the executive sponsor and the operating committee is to ensure that the vision of the platform that has been established and agreed upon is protected. As time passes, changes in business priorities, leadership, and other factors can create major headwinds that cause the path to the vision to drift off-course. It is only with leadership commitment and support that the platform overcomes these obstacles and stays focused on its long-term objectives. Too many roadmaps have been created and left by the wayside as the initial excitement of *phase one* dies down and everyone scatters or becomes tied up in the day-to-day complexities of operations.

It is up to the operating committee to continue to refer to any roadmaps, visions, and guiding principles as the North Star on all decisions, and to raise the alarm when business activities are beginning to stray from that path. Having regular (monthly or quarterly, depending on the footprint and net activities on the platform) business checkpoints to see whether roadmap goals are being achieved and whether the platform is continuing to fulfill the platform vision and effectively supporting the overall business strategy is important when fulfilling this objective. These business review sessions should not focus on the minutiae of individual enhancements, requirements, or defects; instead, they should evaluate whether the platform's current business footprint is aligned with expectations, whether the platform's business customers and users are realizing value from the implementation, and whether any technical, process, or people gaps are affecting value realization, and what resources are needed to close such gaps.

Another important aspect of protecting the platform vision is to reject the next shiny objective that may be placed in front of the operating committee. Stakeholders from various areas of the business may be eager to get on board the platform, and while there are situations where a piece of fruit hangs so low it is highly worthwhile to take a detour to obtain it, it is more likely that deviating from the initial vision may spread resources too thin and reduce time to value across the board. The operating committee must stand ready in such instances to trust and *stick to the plan* and support the platform team against pressure.

Refreshing the platform's vision, strategy, and roadmap

No plan will be perfect as they are always created with imperfect information that has implied and explicitly stated assumptions. The executive sponsor and operating committee should, on a semi-regular (bi-annual or annual) basis, evaluate whether the existing vision, strategy, and roadmap established for the platform are still valid for the business and its strategy. An important aspect of this evaluation is to look at what has been implemented and determine how many more resources to allocate to these areas versus how much to focus on net-new capabilities on the platform. Roadmaps often overemphasize the addition of new capabilities and provide less detail on incremental improvements of existing capabilities. However, during operations, the operating committee should pay close attention to what has already been developed as investments in operations to improve these areas may sometimes have the greatest return on investment with the least risk.

Empowering the platform owner and the platform team to make decisions

While the operating committee plays a key role in establishing and agreeing to the vision and standards of the platform, it is the platform owner and their team who enforce these standards day to day. The platform owner and demand managers must be comfortable with rejecting demands that may not fit the platform vision or are simply unachievable given the available resources. In distributed models, the platform owner should also be empowered to help their platform architects enforce platform technical integrity, which may occasionally mean that certain requirements become more resource-intensive to achieve to accomplish it properly.

Empowering the platform owner is not just a matter of granting them authority – it also means communicating that authority, its limits, and its purpose to those who engage with the platform.

Platform owner

The platform owner leads the platform team and is the key individual in delivering value to the organization from ServiceNow. Let's take a look at their responsibilities.

Communicating and building consensus around the project plan

One of the major jobs of the platform owner is to communicate the vision and plan for the platform to its business customers. This includes socializing the roadmap with the stakeholders, constantly reporting on what is being developed on the platform and why, and how demands and enhancements on the platform are being prioritized. It also includes reporting on whether the platform is encountering any health issues or facing long-term trends (for example, performance issues, a high number of incidents in some part of the platform that may require a more focused problem management effort to address the root cause, or an increase or anticipated increase in HR service requests due to organizational changes such as recent and upcoming mergers).

Communicating the plan clearly and often will help align stakeholder expectations and, in the best case, make business customers partners to the platform and contribute solutions and resources as opposed to being passive consumers of outcomes. When funding, platform direction, and implementation are all centralized under the platform team, this type of communication becomes critical to the platform's success. Without it, business customers can often feel in the dark about when and what they will receive from the platform.

Platform owners should be communicating the plan for the platform in a variety of ways, both one-on-one discussions with the key customer stakeholders and having repeatable assets (a slide deck, a document, and so on) that can easily be distributed to anyone who may be interested.

Using measurement to guide platform decisions

Business decisions, especially ones that can incur large costs, should always be backed up by measurements that allow the risk of failure to be justified. Platform owners will be asked to make many business decisions as part of platform operations, ranging from large (expanding the capabilities of the platform to service a new area of the organization) to small (deciding whether the additional risk incurred by the desired enhancement is worth the business outcome generated to the customer). Ideally, these decisions will be backed up by some sort of measurement. This can be easier said than done, and entire books exist about organizational KPIs and how to measure and model the risk of decisions properly. While we will not cover this topic in detail, we will provide a few guidelines for success regarding this topic:

* First, it is important to be aware of whether a decision needs to be made. A decision should have more than one viable option and there should be some cost associated with each decision. There will also be an element of uncertainty with each decision; if you are certain about each, then it should be trivial to understand which is better. Similarly, if each option is equivalent to another, there should be absolutely no hesitation in picking an option at random.

* Second, models and measurements should only be created and made if the possible result of that measurement will lead to a different decision. If you cannot establish ahead of collecting and measuring the desired data how the results of that data will cause you to choose between one or more options, it is unlikely that the measurement will be worth it. This rule also means that many of the complex or busy "dashboards" built by various business groups (including those by the ServiceNow platform team!) are irrelevant when it comes to generating business decisions. Instead of focusing on collecting more data that may not be relevant, the platform owner and their team (and any other business leader) should focus on what kind of decision they are looking to make and what kind of data will enable them to decide upon one decision or another.

- Third, never be afraid of treating data collection and measurement as a series of experiments instead of a single long-running initiative. Collect the right data and measure the right information for a given decision and move on after the decision has been made. There will rarely be long-running metrics that will continuously provide answers to inform continuous decisions (though there are some, such as looking at sudden and substantial changes in certain steady-state metrics that can drive decisions on problem investigation).

Advising the operating committee on what is needed to execute the platform's vision

The platform owner is likely to be the most senior member of the operation team and has direct and deep knowledge of the business capabilities of the platform. When working with the operating committee, it is the platform owner's responsibility to provide information to the rest of the committee about aspects of the platform that they may not be familiar with. This includes, among other things, the platform's capabilities, limitations, and costs. While the platform owner may also not be an expert on each of these subjects, they must gather a team of technical leads, product owners, and architects that can provide them with this knowledge.

With this knowledge, the platform owner's role regarding the operating committee is to ask for the right support to effectively execute the platform vision that has been established. If the direction and implementation capabilities of the platform are all centralized under the platform owner, then there will be an increased demand for the platform's owner's ability to create the right plan and obtain the right resources to be able to establish and then execute the vision.

Business analyst/product owner

The business analyst, or product owner, serves as the individual translating the needs of the business into platform requirements and designs. In terms of the platform team, they should take care of various aspects. Let's take a look.

Understanding the needs of the customers and translating them into requirements

The business analyst/product owner's role is to translate the needs of the customer into a *product* that serves those needs. In the case of the platform operations team, that *product* is the set of changes on the platform that must be made to meet those requirements. The product owner and business analyst aspire to know the business of the customer better than they may even know themselves. With a deep level of insight into the customer and with knowledge of the platform's capabilities, an exceptional product owner can create requirements on behalf of their customer that address or exceed the customer's expectations, as well as have requirements that will be well-aligned with the platform's capabilities and avoid playing into the platform's limitations.

The knowledge of the customer not only applies to creating great requirements but also enables the platform/product owner to be able to ruthlessly prioritize the backlog of priorities to deliver the right functionality to the customer in the right order and at the right time.

Understanding the customer's needs also translates into communicating and managing the expectations of the customer so that they are aware of how capabilities will grow to meet their needs, both present and future. Therefore, the product owner acts partly as a relationship manager between the platform team and the customers of the platform and should help the platform owner understand where customer needs are being met and where there are deficiencies (either in perception or reality).

Interfacing with the technical team

The product owner must be able to communicate the functional requirements of the customer to the technical team so that they can be decomposed into a set of technical designs and platform configurations that allow the platform to meet those functional requirements. In a typical agile development environment, this means that the product owner must create artifacts that enable the functional requirements to be easily digested. They must walk through these artifacts with the development team to confirm that they are understood and estimable. Typical artifacts that are used to communicate the functional requirements to the developers include wireframes, mock-ups, user stories, and acceptance criteria. It is important to note that the format of the communication is not as important as being able to communicate the information concisely and clearly. Functional requirements communicated in this way must be somehow functionally verifiable after the configuration has been completed. Requirements not explicitly stated and not explicitly verifiable will be prone to "defects" as differences in interpretation between individuals cause discrepancies in what the product owner expects and what the developer has built. A small but common example of this would be positioning the user interface elements on a form. If no mock-up is provided and no details are specified for where a user interface element is to be positioned on the form, it is highly likely that a requirement such as *there should be a button to cancel the request* could result in the positioning of the button being different than what is expected by the product owner.

When developing requirements and interfacing with the technical team, the product owner needs to avoid providing technical solutions to the developers as much as possible, even if the product owner has enough technical understanding to know how a particular functional requirement may be addressed. Technical teams (in consultation with the architect) should always have the last say in how a particular functionality should be implemented. This provides the team with the greatest level of flexibility in meeting the functional requirement using the best technical options available. The assumption here is that the functional resource will not be as technically capable as the technical team, and even if they were, the functional resource will know fewer of the details of the platform, so they will be less qualified to make technical judgments.

Demand managers

As the organization stands up the ServiceNow platform, there will be no shortage of demands from stakeholders on how the platform can be made better. The demand manager's job is to make sure the great ideas that are aligned with the platform's strategy are surfaced, and frivolous requests are deprioritized. Great demand managers are constantly working on various aspects. Let's take a look.

Managing the pipeline of incoming demands and improving the overall quality of demands over time

The key role of the demand manager is to manage the queue of demand from the product owner/business customers. Demand managers should recognize low-quality demands and reject them outright. For demands without enough detail but with high potential, demand managers should request the demand submitter to provide the required information to help the demand manager qualify the demand.

Demand managers differ from product owners in that demand managers work with a broader array of stakeholders and are more likely to encounter requests that are not qualified or are not aligned directly with the platform's strategy and vision.

Beyond qualifying demands, demand managers also help shepherd the demand through the demand management process within the organization. This commonly only applies to demands that require budgets beyond the typical operational budget (you would not normally go through the demand management process for small enhancements that may be passed directly to the product owner).

Platform architect

The platform architect is not just a senior technical resource. At their best, a platform architect is a person deeply connected to the business and the entire footprint of the organization's technologies who can help guide the technical design of the most crucial parts of the platform. The platform architect should identify and produce the shared data and architecture of the platform to be communicated by all technical teams.

Shared data and architecture are platform configurations that are consumed by specific modules and other configurations to accomplish business outcomes in a specific way. Once again, the CMDB and the **Common Service Data Model** (**CSDM**) are notable examples of shared data and architecture components. Regardless of the module, when the CMDB data is created or consumed, developers should follow common conventions in terms of which tables the CI data should go to and what attributes should be used for what purposes.

When the shared architecture is designed by the architect, they must also communicate how this shared architecture should be used to all developers (often via developer leads). This communication does not need to be verbal – in fact, even for small-scale teams, there is a significant value in documenting the architecture design, the reasoning, and examples of how to utilize the shared architecture both in terms of efficiency of communication and in the effectiveness of knowledge transfer for new and external team members.

Sometimes, it will not be obvious to an individual developer or product owner that a particular requirement calls for a shared architecture design. This is where the architect must leverage their experience and visibility across streams to identify these situations and work with the product owner and technical teams to incorporate the development of the shared architecture component as part of the requirement. This step is particularly important as not identifying the right shared architecture opportunities may result in development creating many inefficient point solutions on the platform that could have been simplified and consolidated into a single design. Nevertheless, because this process can never be flawless, an additional task of the architect is to identify situations where multiple point solutions should be merged as a single shared architecture component. They must work with the demand manager and/or product owner accordingly to allocate the correct time and resources to perform this consolidation at the right time.

Summary

In this chapter, we covered the key aspects of operating the platform that the organization should consider to improve the chances of obtaining a positive return on investment.

When your organization selects ServiceNow, make sure that you have considered who the immediate customers are that the platform has been built to serve, how the platform will fit within the broader technology landscape of the organization, what the operating model for the platform is, including demand management, implementation, and ongoing maintenance, and how to create the right team in service of the platform and the business to generate long-term value.

In the next chapter, we will move away from the people and processes side of the equation and highlight a cutting-edge feature of the platform that can be integral to value realization: artificial intelligence.

10

Artificial Intelligence in ServiceNow

This chapter will cover one of the most powerful areas of the ServiceNow platform: its **artificial intelligence (AI)** and **machine learning (ML)** capabilities. The ServiceNow platform includes diverse AI features including predictive intelligence, AI search, Health Log Analytics, and Virtual Agent with Natural Language Understanding. This chapter will focus on predictive intelligence.

This chapter's goal is to make the predictive intelligence AI features and their use and implementation accessible to as many people in the ServiceNow community as possible. We'll start by providing a brief introduction to the relevant concepts in AI that are directly applicable to the ServiceNow capabilities. This will help you build a basic understanding of how you can expect the system to behave.

With that foundation in place, we'll cover the key AI/ML capabilities of ServiceNow, as well as their application, and provide a framework. This will help you implement these features so that they have an impact on your organization.

In this chapter, we will cover the following topics:

- Understanding AI and ML
- The AI/ML capabilities of the ServiceNow platform
- Classification framework
- Regression framework
- Similarity framework
- Clustering framework

By the end of this chapter, you'll be able to understand, explain, and implement the most common use cases of ServiceNow's AI features.

Understanding AI and ML

When implementing AI features, it's incredibly useful to have an understanding of what the system is doing so you know why you're getting certain results and how those results will change in response to different actions you can take. A full introduction to AI requires at least an entire book, which would likely be out of date within months of publication. Therefore, we'll cover the bare minimum needed to interact with ServiceNow AI deployments here. We will try and focus on enduring concepts rather than the latest trends.

What is AI?

Broadly speaking, AI is the ability of artificial systems to understand, process, or make decisions based on data provided to the system. AI is often used synonymously with ML, but ML is a subset of AI and some AI systems do not incorporate a learning component.

The difference between AI and ML

Even a simple set of scripted conditions is AI in a sense, and the earliest AI systems essentially consisted of lots of rules coded in by people. Unfortunately, once you start writing down rules, it quickly becomes apparent that, in most cases, the world is too complex and dynamic for people to keep up by writing all the rules manually.

In response to this, a sub-field of AI called ML has been created that allows people to provide computers with lots of data and then have the computers process that data to create their own rules that best fit the past data. This works well because the rate at which people can generate data is often much faster than the rate at which we can develop consistent, generalizable rules about the way the world works.

Typically, these sets of rules are called **models**, but ServiceNow refers to them as **solutions** in some cases. You can think of a model as a formula that relates a set of inputs in some standard format to an output of some kind. One kind of model might take a string of words and output a number representing the sentiment (positive or negative), while another model might take in an image and output a word to describe the contents of the image.

Again, you could, in theory, come up with a model manually but the central feature of ML is that the models are created by first defining a structure (how many variables and how those variables are related to each other) and then running some process to populate the variables in a way that best describes the training data. Some other nuances and techniques are used to prevent the models from overfitting to the training data in a way that would make them less useful. However, it's enough to know that the models are populated mostly by using training data and that a good ML model will be able to predict future values using the information learned from past data.

Metrics and tradeoffs in ML models

A wide variety of metrics are used in ML research and development, but we'll focus on the ones ServiceNow uses in their platform – **coverage** and **precision**. Coverage refers to the estimated proportion of future records that will have predictions made, while precision refers to the estimated proportion of those records for which predictions were made where the said predictions ended up being correct.

There are usually tradeoffs between different metrics, and understanding these tradeoffs is often essential for tuning a ServiceNow model for maximum impact. To manage these tradeoffs, it is important to understand that, in a business context, a model can give correct answers, incorrect answers, or no answer at all. You may have to make decisions such as whether you should increase the number of correct answers per wrong answer at the cost of giving fewer answers overall.

Imagine, for example, a five-question test where you would score 1 point for each correct answer, -5 points for each incorrect answer, and 0 points for a skipped question. You'd probably only want to answer the questions you were most sure about and skip the rest. If you were to write the same test and you got +5 points for a correct answer and -1 point for an incorrect answer, you'd probably be more willing to risk a wrong answer because the penalty is lower.

The same is true in the ServiceNow domain. If a customer creates a case about an issue, it would be quite acceptable to have a model recommend a possible knowledge base article to them 90% of the time, even if it was, on average, only 50% likely to solve the issue. This is because the cost of showing a user a non-helpful article is very small compared to the value of solving their issue on the spot.

In contrast, if you're using a model to route tickets, you'd be happier to have a model that only gave an answer 50% of the time but was more than 90% correct when it did make a prediction. It's ok for your model to "admit" to not knowing the answer but you don't want the tasks sent to an incorrect group, only to languish in their queue until they re-assign it again.

These two situations illustrate cases of different priorities since the confusion created by an incorrect knowledge article is relatively small but the confusion caused by an incorrect routing can be much larger.

It's worth noting that not all ServiceNow predictive intelligence features can assess their own precision metrics. For example, when looking for similar records using the similarity framework, you may need a person to manually check if a given answer is a *good answer* based on the input data.

The role of data in ML

As you may have realized by now, data is an essential component of an ML solution. The *learning* part of ML involves processing data to create a better model. While different types of models approach this in different ways, a common trend is that the quality of the output of the model will be no greater than the quality of the input data, even if the best model type is selected.

In a typical ML project, you will have some part of your historical data that you use for training, often called the **training set**, as well as two parts of that historical data that you use to assess how good your models are, known as the **validation set** and the **test set**. ServiceNow handles these splits for you and uses them to calculate the estimated metrics for your solutions as you train them, so you don't need to worry about these splits.

Data comes in many forms, whether it is images, words, or numbers. However, in most AI models, the input data needs to be converted into lists of numbers in some way; those numbers are what is processed by the AI. Converting inputs into numbers is particularly important when we talk about text data. Fortunately, a great deal of research has been done into processing natural language text, such as what you might see in an incident short description, so there are tools that ServiceNow can use to make this process easier.

In **Natural Language Processing** (**NLP**), converting text input into numbers is often separated from the task of doing something useful with those numbers. We call the first part of the process **embedding** the text. This can mean taking a word such as *cat* and turning it into a list of numbers such as [0.2309, 0.2828, 0.6318, …, -0.7149]. Once the words are in numeric (also known as vector) form, you can use those numbers to do a variety of things, such as classification, clustering, and more. Each type of data you find in ServiceNow has some conversion that can be applied to put it into a numerical form. The algorithms will use those numerical representations for learning purposes.

Once the data is in a form that is suitable for an ML model to operate on, you can apply a variety of different algorithms to that data. In the next section, we'll look at some examples of these algorithms.

Algorithms in ML

To understand how ServiceNow's models can be expected to work, let's look at some very basic algorithms that can solve specific problems. ML models might use many inputs or a complex input, such as a field of text, but the learning process results in a model that minimizes the mistakes it makes on the training dataset with the assumption that how the model does in the training will be an indicator of how it will do with real-world data. First, let's predict a number based on some set of inputs. This type of task is called **regression** and focuses on the ability to correctly predict a number, given input. One common example of a regression problem is that of predicting a house's price when given some inputs such as size, age, and the number of rooms.

Let's look at a visual example. Assuming you wanted to predict house prices based on the number of rooms, you might plot the price against different numbers of bedrooms and then train a model to fit this data as well as possible. See the following figure for an example:

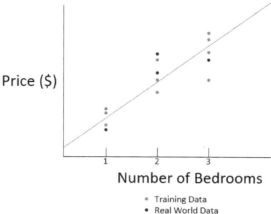

Figure 10.1 – ML works when the real-world and training data are similar

The red line indicates a possible model that has been trained to *fit* the training data to minimize the total *error*. As you can see, the real data seems fairly similar to the training data, so this model does a pretty decent job. However, if you used data from one city as your training data and then tried to predict house prices in a different city, you might end up with something similar to the following:

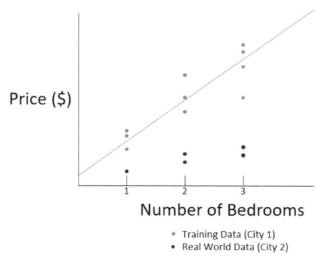

Figure 10.2 – When the training and real data are very different, ML will
do well on the training data but poorly on the real data

In this example, you can see that the training data and the real-world data are unlikely to have come from the same underlying distribution, which means that no matter how well your trained model fits the training data, it's not going to do a good job on the real data. Therefore, it's important to be cautious any time you want to undertake an ML project and you don't have access to a good quality training dataset that is representative of your production data.

The AI/ML capabilities of the ServiceNow platform

This section will focus on the different AI and ML capabilities within the ServiceNow platform. For each technology, we'll cover not only the features but also examples of use cases where these features add value. By the end of this section, you should have a good overview of the capabilities offered, as well as how they can be applied in practice within your organization.

Predictive intelligence

ServiceNow's predictive intelligence is a collection of fundamental ML features that handle similarity search, classifications, regression, and clustering, along with the supporting features that enable customers to train, manage, and support these features. Predictive intelligence underpins some of ServiceNow's features, such as *Major incident Workbench's Similar incidents* feature, but these capabilities are also available via a set of developer APIs so that they can address use cases specific to your business as well.

Within predictive intelligence, you'll find four types of solution definitions, each targeting a specific type of problem. Later in this chapter, we'll cover the steps to bring predictive intelligence to your company across these four capabilities, but first, let's go through a hypothetical example of a predictive intelligence journey to give you a rough understanding of some of the key elements of this process.

Practical example – automating an incident routing rule

The best way to understand the difference between AI and ML is with a simple toy example that is representative of the journey many customers find themselves on.

Let's imagine an incident has been logged in ServiceNow and, immediately, the company's on-call service desk agent gets an email notification asking them to route the incident to the hardware or software team. Within 30 minutes or so, they will read the incident description and make a judgment call based on their knowledge of technology issues at this company and choose to either send the ticket to the hardware or software team.

After a while, the company leadership questioned the necessity of a trained specialist looking at every ticket and instead asked the service desk team lead to create an automatic process to route the tickets. The lead discussed this with their ServiceNow architect and together, they wrote down a set of rules to route tickets and then sent it over to the ServiceNow development team to implement.

The dev team looked over the rules and decided to build ServiceNow assignment rules based on these rules that look for certain keywords, such as *Laptop* or *Microsoft Word*. Then, depending on the matches, they send them to the hardware or software teams. If there are no matches, then the ticket will be routed to the software team because they generally have more capacity than the hardware team.

With this new process in place, the service desk team had time to spend on higher-value tasks; the tickets were mostly routed very quickly but there was an increase in routing mistakes.

After a while, users start to complain that tickets related to the desktop interface for the recently implemented ERP system keep getting sent to the hardware team rather than the software team. The process owner recognized that while assignment rules and branching logic like this can cover some static rules, what they needed was a process that could learn and adapt to change – in other words, an ML solution. The process owner recognized that an ML system would allow the system to adapt without keeping the dev team busy coding up new rules each time something in the environment changes. Fortunately, ServiceNow's predictive intelligence solution provides several capabilities that cover some of the most common ML use cases.

By enabling ServiceNow's predictive intelligence solution for classification, they were able to train a classification solution (which is a type of ML model) so that it could use data about where past tickets were and notice patterns in the data that the service desk probably used subconsciously but didn't consider important when they were writing rules. For example, words such as *battery* and *USB* occur with relatively high frequency in hardware issues than software issues, and words such as *desktop* and *computer* don't tell you as much about where to route tickets as you might think.

The advantage of ML is that with enough input data, the system will learn many different factors and that these factors can continue to be updated over time without human input. These patterns can be used by ServiceNow to decide where to send the incident and recognize when it is not sure enough to make an automated judgment so that human judgment can be made instead.

The new system worked well but still didn't quite live up to the expectations the process owner had for their incident routing. Due to this, they completed some error analysis by looking at cases where the predictive intelligence features didn't provide the results they expected. In some cases, the errors were understandable (for example, long rambling incident descriptions describing a multitude of issues), but looking at the data showed that the system would generally be correct when it assigned tickets to the hardware team but that it would often send hardware tickets to the software team as well.

This result got the process owner thinking back to the old rules-based model and they realized that they had a systematic issue in the data where tickets had been misclassified as software issues by the old rules-based system. Based on this analysis, the service desk team was asked to use some of the time that this new automated process saved them to find examples of recent software tickets and to correct them, as well as any future tickets.

This data-cleaning process finally allowed the company to realize the full value of automation in ticket routing. While the system didn't always get the routing perfectly correct, it would get most of the tickets to the right people far more quickly than a manual process and far more accurately than the rules-based approach.

Key lessons

This short case study illustrates several very useful concepts that you can apply to your implementation of predictive intelligence:

- This was a great use case because there was often a clear right answer for each classification task and there was already a team responsible for making those judgments. This type of use case is far more likely to succeed than use cases where the inputs, outputs, or decisions cannot be articulated clearly.

- The fact that this company had people doing the routing at the start of their journey allowed this team to qualify another important AI metric, known as **human-level performance**. Today, a trained person is the most intelligent system we know of, so most AI systems are aiming to get close to human-level performance, which is the expected performance of an unrushed human expert with access to the same data. Most systems aim to reach human-level performance and it is rare for AI systems to exceed human-level performance in terms of accuracy, so you can consider this level of performance as a likely upper bound on how good the system could be under ideal conditions.

- Implementing and testing a basic set of rules before moving to an AI solution provides what is called a performance **baseline**, which can be used as a likely lower bound on the performance of an AI system. If your AI system is performing more poorly than a set of simple rules, it's unlikely that it has been properly selected, implemented, and tuned. We'll cover the use of human-level performance and baseline performance to assess an AI use case later in this chapter but for now, knowing how good the baseline is gives you a sense of how hard the problem is.

- Implementing the predictive intelligence features required a substantial dataset of past incidents to train the solution. In general, ML is not a capability that can be implemented without data because there will be nothing for the ML system to learn from. At the same time, issues in that data can impact the quality of your system's results, and the patterns of error in the training data were ultimately reflected in the results the classification solution provided.

- Correcting the data issues in both the past training data and new records allowed the company to improve the performance of the model but required time and effort from the service desk. While teams often focus on the reduction in effort from implementing ML, it's important to recognize that new tasks will need to be carried out to keep the system performing well over time.

- The final value proposition for this case study was not that the system outperformed the service desk from the point of view of precision, but rather that it was nearly as precise but far faster than a human and roughly as fast but far more precise than the rule-based models.

Now that you have a conceptual understanding of the key concepts and have an idea of how an ML deployment might progress, let's have a look at some of the specific features ServiceNow offers, as well as how they can be applied to drive value for your organization. We'll start with the feature that this example was built on: the classification framework.

Classification framework

Classification is the task of taking a set of input data, including text or other fields, and using it to output a category, assignment, or other value from a set list of options. In traditional enterprise workflows, simple classification decisions are made with static rules, while more complex classifications rely on expert judgment. ServiceNow's classification solutions allow the systems to make more complex judgments in an automated way by leveraging information about how data was categorized in the past.

Classification solutions generally outperform simple rules when there are complex relationships between multiple input factors or when there are many examples of high-quality training data for a specific problem. In contrast to this, in a common use case such as setting a priority based on impact and urgency, the relationships between impact and urgency are typically clear and can be captured in a simple 3x3, 4x4, or 5x5 data table. In these cases, you don't need ML to create an accurate classification because simple rules already work well. If you take the task of predicting the category of an incident from input data, including the department of the user, the text of the incident, and the country of the user, you start to get complex relationships that are not possible to fully document, much less implement in code.

Assessing the applicability of the classification framework

Once you have an idea where a classification solution could be applied, you can use the following tests to determine if this is the right type of solution for your use case:

- Is the problem more complex than you could solve with a simple set of rules or a lookup table? For simple problems, using a classification solution is needlessly complicated.

- Is the problem you're solving valuable enough to justify managing a classification solution over time? Typically, you'll want to target higher-volume use cases where the efficiencies gained will outweigh the costs to support and tune the solutions.

- Are the output values (the ones you want to predict) a choice or reference list with a limited number of options? The more classes you have, the harder it is for the model to predict useful values.

- Do you have enough high-quality data to train a classification solution? You will need at least 10,000 records in aggregate with 30 or more examples of each class. However, ServiceNow recommends over 30,000 records for good results. Keep in mind that errors in your training data will make the model's predictions worse.

- If a mistake is made in the classification, is there a way to correct that error before unacceptable harm occurs? Classification solutions typically have some error rate, so they should not be used if mistakes would be unacceptable (for example, in safety-critical applications).

If you can answer yes to these questions, a classification solution is likely worth trying for your use case. Now, let's see what it takes to configure a classification solution.

Getting started with the classification framework

Once you've applied the criteria in the previous section to determine that your use case seems to be a good fit for the classification framework, you can train your first classification solution. The technical steps are available in the ServiceNow documentation and are subject to change with new releases, so we'll focus on the key decisions that need to be made and how to think about them.

First, you will want to create a classification definition, which you can do by going to the left menu and selecting **Predictive Intelligence | Classification | Solution Definitions**:

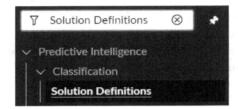

Figure 10.3 – Menu location of Solution Definitions

The classification definition form walks you through the steps involved, but we'll provide some supplementary details for how to think about each of these steps:

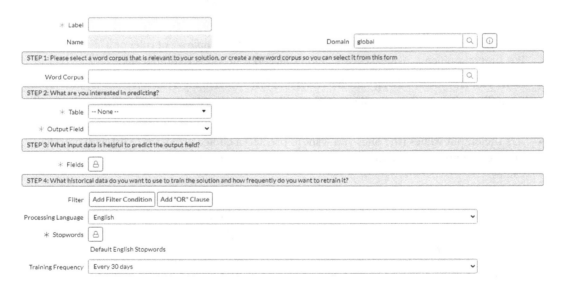

Figure 13.5 – Classification definition form

The first thing you must do is create a descriptive label and review the name and, possibly, the domain of the definition. The label is up to you and does not impact the functionality of the solution definition. Let's look at each step one by one.

Step 1 – word corpus

If you plan to include any string field (*string* is the technical name for text fields) in your model, you should create or select a word corpus. Remember earlier in this chapter when we discussed *embedding* words as vectors or lists or numbers? The word corpus is used to complete this process. It's important to note that you can reuse a word corpus between use cases and that the word corpus can include tables other than the one you're predicting.

A good word corpus should contain a lot of text that is relevant to your company's use cases. If you don't have a lot of good text data, you can use the **Pre trained** option, which provides decent performance on general language but won't capture the details of your company's systems and nomenclature. If you include a text field but don't create a word corpus, the system will try to automatically create one. However, it's better to intentionally select your options for this field if you want to use any text processing capabilities.

Step 2 – output fields

After selecting a word corpus, you'll want to identify the table and field that you are aiming to predict. If you want to predict multiple field values, you should create a classification definition for each one. Remember that you want to choose a field with a limited set of options. ServiceNow prevents you from selecting text fields as the target because the classification algorithms they use aren't useful for predicting freeform text. It's also important to choose a field that has a single logically correct value. For example, you could reasonably ask the solution to choose an "assignment group" but choosing an "assigned to" individual requires context surrounding availability, skills, and other data that is likely not present in the input record.

Step 3 – input fields

Here, you'll want to select a set of fields that can be used to predict the output values. When selecting these fields, you should consider at least the following factors:

- **Input sequence**: The fields being used to drive predictions should be populated before the output fields. For example, asking to predict an assignment group using a closure code wouldn't work because the closure code is populated later in the workflow.

- **Relevance**: Ideally, you want to choose fields that have a clear causal relationship with the output; for example, the description field of an incident is very useful for determining the categorization that should be assigned. A good way to think of this is to ask what information from the record a person would want to use if they were doing the prediction themselves.

- **Stability**: A field that is consistent across the life cycle of your record is a much better predictor than one that changes between the time of prediction and closure. For example, the state of a task is a very bad input field because your training data likely consists of mostly closed tasks, while the tasks you want to predict based on will likely all be open.

It can be useful to experiment with different input fields and see how they impact the model output.

Step 4 – training data and frequency

In this step, you can restrict the set of data that you want to use for training. This is useful when not all the data in the table from *step 2* is useful for making predictions. You will likely want to exclude data that is too old (more recent data is more likely to be relevant than older data) or that is still in progress (for example, tickets still waiting for assignment).

The **Stop Words** field allows you to select a list of stop words that should be ignored, often because they are very common or don't provide useful information about the ticket. Stop words are most useful in fine-tuning your results. So, if you're operating in English, it's usually alright to use the *default English stop words* as a starting point and then add your own company-specific stop words after looking at some of the results where the classification solution didn't perform as expected.

Finally, the **Training Frequency** field allows you to set an interval for your solution to be retrained automatically. This helps you incorporate information from new records and keeps your model from becoming stale. Once you've filled in all the fields, you can send your solution for training using the **Submit and Train** button.

Evaluating a trained solution

When you train a solution definition, the system will create and train a new ML solution for classification, which includes detailed information about how the classification solution is expected to perform. The amount of time it takes to train the solution is dependent on multiple factors, including the size of your training dataset. However, once the state is set to **Solution Complete** by the training process, you'll be ready to evaluate and tune your model's performance. To do this, you'll have to rely on three main metrics that the ServiceNow platform calculates for you.

Precision

Precision is the percentage of the model's predictions that we expect to be correct. A precision of 90 means that 90% of the predictions were correct. Precision is a useful metric if the simple ratio of correct classifications to incorrect classifications is very important to you, but always remember that surprising simple rules can often provide high precision values without generating a lot of value. For example, if you meet a stranger in the street, you can predict with a precision of 99.7 that today is not their birthday simply by always guessing it's not their birthday. This doesn't mean you're *good* at guessing birthdays but rather that high precision can be achieved with a simple rule.

Coverage

Coverage is the percentage of records for which the model is confident enough to make a prediction. A coverage of 80 means that the model predicted 80% of records. Coverage is important because admitting that you have insufficient information to make a confident prediction prevents the model from making avoidable errors. This is important because the cost of a bad prediction can often be much higher than the cost of making no predictions at all.

Recall

The weighted average of the recall of each class. This definition of recall is the number of times the solution predicted the correct class out of the total number of occurrences of that class, aggregated across all classes. This is a metric that can help you understand how well the model is doing at predicting the different classes, but this metric is best used on a class-by-class basis. (This information is available on the **Class Confidence** tab.)

There is an inherent tradeoff between precision and coverage, which we addressed earlier in this chapter. Now, you can apply the concepts from that discussion here by selecting a **target metric** and indicating a value for the solution you want to try and achieve. Applying the value usually won't get you to the exact target value but will bring the solution closer to that point.

The **Test Solution** tab allows you to input various test values and evaluate the predicted outputs of the model directly. You also have the option to run the test against a subset of the target table to assess the predictions there. This can be useful for checking the model's performance on very recent data.

Related lists

There are two important related lists for the classification solution: **class confidence** and **excluded classes**.

The **Excluded Classes** tab lists those classes that you didn't have enough training data for, which means they won't be predicted by the model. Remember that you need at least 30 data points for a class to be predicted.

The **Class Confidence** tab shows you the precision, coverage, and recall for each class, as well as the distribution of classes in your data. It's important to look at this because if your data is very imbalanced (that is, some classes occur much more frequently than others), the model could have good aggregate metrics by simply doing well on one or two of the classes. Note that a model that is good at predicting a smaller set of classes can still add business value if the cost of mislabeling a record is relatively low. Classification is just one of the capabilities that ServiceNow's predictive intelligence features enable. Next, we'll look at regression, which predicts numeric values instead of a class.

Regression framework

When you consider the different types of fields in ServiceNow, you might notice that not all fields lend themselves to a classification approach. In addition to assigning categorical values, you may also want to predict numeric field values such as durations, sizes, effort levels, or other measures that vary along some numeric scale. Regression solutions have some distinctive characteristics, which make them very useful as an additional tool for automating processes. To understand these benefits, it is necessary to think about the types of data that we would be predicting and how a person would estimate the number based on inputs. Let's use the example of estimating the hours that should be estimated for a user story based on some given input text and an assigned team.

The first thing to think about in this situation is whether we should hope for a precisely correct answer each time. If you estimated 30 hours and it turned out to be 28 or 32, you'd probably still think it was a decent estimate; regression solutions approach the problem in the same way. Training a regression solution minimizes how wrong the model is without really expecting that it will get the right answer each time. Instead of an exact value that can be nearly impossible to guess, it may be more useful to think about things in terms of a range of likely values. A person might say, for example, *That feature will take between 20 and 30 hours to implement*. A regression model in ServiceNow can do the same thing when it provides a range for the expected value of a given value.

Metrics for regression

This understanding allows us to talk about the metrics that are used to assess a regression model. Remember that these metrics are about getting close to the actual value, not getting a precisely correct answer each time.

The metrics for a regression solution will be covered in this section.

Mean absolute error (MAE)

On average, how far from the correct values are the predictions? This is a measure across your whole dataset, so it doesn't tell you about how bad your worst case is or how good your best-case estimates are. If you had two records where the correct value is 10, you could get a **mean absolute error** (**MAE**) of 5 by predicting 10 and 0 or by predicting 5 and 15. It's important to note that these are absolute errors in that they are *absolute values*, so an overestimate doesn't cancel out a later underestimate. What matters is how wrong the prediction was, not in which direction. The units for MAE are the same as the units for the field you're trying to predict. So, if you are predicting hours, then the MAE would be in hours as well.

Symmetric mean absolute percentage error (SMAPE)

Symmetric mean absolute percentage error (**SMAPE**) is important because MAE is not always a good way to look at errors; for example, predicting a value of 1,005 when the true value is 1,000 might be acceptable (MAE=5), but predicting 5.25 when the true value is 0.25 (also MAE=5) is not nearly as good. To get around this problem, SMAPE, a modification of MAE where you divide by a special value after getting the absolute error, can be used. To avoid dividing by zero and to make sure we can always give an error value between 0 and 100, the system calculates this special value as the average of the prediction and the actual value. This gives you an answer between 0 and 100 but as a result, the answer will no longer be in the same units as the predicted fields.

Range accuracy

We mentioned that regression models not only predict a value that should be close to the true value but also a range (lower and upper bounds between which the value is expected to fall). The range accuracy is quite simply the estimated percentage of predictions that will fall within the range predicted. Predicted ranges are very useful when planning because it gives you some idea of what the likely outcomes are.

Average interval width

An excellent range accuracy can be achieved simply by providing a very wide range. How much time will that feature take to develop? If you answer between 1 minute and 42,000 years, you're probably right, but your answer won't be very useful. The average interval width provides a measure of the average distance between the bottom of the bound and the top of the bound. If this distance is smaller, this points to a more useful model if range accuracy is still usefully high.

You'll notice a consistent trend throughout this chapter, which is the tradeoffs between metrics; this tradeoff is inherent to the ML field. Your job, as the person deploying, sponsoring, or analyzing these results, is to consider the business implications of these tradeoffs given their place in your processes and operational context.

These metrics will allow you to assess how well your model works. In the next section, we'll look at the importance of data quality in training models to get good results on these metrics.

Input data quality

When you go to train a regression model, selecting the input dataset is very important. Due to the fact that errors are averaged, a population of easily predicted values in your training data can have a big impact. Regression models require at least 10,000 records to train, which means that it's tempting to include records with no value or the default value to create a large enough dataset. The problem with this is that your model will be able to predict some number close to the default value for each record and artificially deflate the MAE and other metrics. Instead, consider strategies such as labeling some of the records manually. Even if this will take a few person-days of effort, this will lead to far more accurate results and could make the difference between the success and failure of your use case. Most teams typically under-invest in assessing and improving their data when working on ML use cases because it's not the most engaging work or because the task can seem insurmountable when you're talking about thousands of records.

Improving data quality

There are a few strategies that you can use to help assess and reduce the efforts to improve your training dataset.

Strategy 1 – assess the true size of the effort

Consider working through a small set of 100 records manually and seeing how long it takes. Then, use that to estimate how long it would take to label or validate your entire dataset.

Strategy 2 – validate a representative sample

If you're unsure of the input data quality but are not able to justify evaluating every record by hand, you might consider evaluating 5% or 10% of the records selected at random. This will give you a very good estimate of the overall data quality at a fraction of the cost.

Strategy 3 – develop the value case for good data

If you validate some or all of your data but find that the quality of data is poor, this can strengthen the value case for your ML use case. It indicates a gap in the current process and an opportunity for even greater improvement relative to the current state.

Regression and classification models both have a common theme of predicting field values. The next two model types don't specifically predict a field value, but they can both be very useful.

Similarity framework

The ServiceNow similarity framework allows you to find new records that are similar to a given input record. This could help a support agent to find past customer cases similar to the case they are working on so that they can reuse knowledge or resolution details. It can also help prevent entering duplicate data or correlating records coming in, such as incidents related to an active major incident.

The similarity framework differs from the prior two predictive models because it will find entire records and not just a single field value.

A useful trick when employing a similarity search model is to use data values from a highly similar past data record to populate your current record. This allows a good similarity model to act as a classification or regression model. This approach is not without its drawbacks however and ServiceNow is not set up to calculate metrics for that use. In general, when using similarity models, it can be hard to assess metrics because determining whether a person would consider records to be similar requires human inspection.

Assessing the similarity framework's performance

Because ServiceNow doesn't provide built-in metrics that assess similarity model results, you need to employ different methods to confirm the usefulness of similarity data.

ServiceNow allows you to manually review similar record examples. This is a quick and often useful way to get a sense of what the model considers similar. It's quick and easy to use and should always be your first step in looking at the output of a similarity framework solution.

To complement this list, there's a more rigorous but also more time-intensive approach that can be used. You may choose to adapt some of this process to meet your needs, but the general structure is given here:

1. Before training your model, set aside a small set of test records (say, 50 to 100). Be sure to exclude these records from your training dataset for your similarity solution.

2. Train your model on the remainder of your training dataset.

3. On the solution version page, there's an option to test records, run your sample records through this testing feature to identify the most similar records, and use a similarity score for each match. Record these results.

4. Take these results and give them to two to three different people who are knowledgeable about the tickets and ask them whether they would consider the records to be similar. (Giving the records to multiple people allows you to minimize the subjectivity of the assessment.)

5. At this point, you can construct a table of records that specifies their similarity score and human-assessed relevance. Rank the entries in the table by similarity score with the highest similarity records first. It is likely but not certain that records higher in the table will have better human scores. This table will allow you to set a similarity score threshold that gives you acceptable coverage while providing results people consider useful.

Let's look at an example that consists of 10 test records assessed by two human reviewers:

Test Case #	Similarity Score	Human Assessment
1	87	1
9	86	1
6	86	0.5
2	86	1
5	83	0.5
10	76	0.5
7	73	0
8	72	0.5
3	72	0
4	68	0

Table 10.1 – Example similarity scoring table

In this example, records with a similarity score of 75 or higher mostly seem to be useful to human validators. In this case, setting your similarity threshold to around 75 would make sense and provide around a 75% usefulness rate and 60% coverage.

Using these tools, the ServiceNow similarity framework can be incredibly useful for providing relevant current or historical data in the context of work being done on the platform.

Another example of the similarity framework is when using a pre-trained word corpus. Here, it's possible to create a relatively good model with very little training data. Similarity is the only one of ServiceNow's predictive intelligence features that can be turned on from day 1 of your ServiceNow journey, even without historical data, though you should expect better results with high-quality training data.

The final predictive intelligence model type is a sort of bulk similarity model, and is called the clustering framework because it clusters together groups of similar records for analysis.

Clustering framework

Clustering is a tool for exploring data and detecting patterns that are not apparent to people looking at individual records. Like similarity solutions, a clustering solution quantifies the degree to which records are like one another. The difference is that while similarity search applies to one record at a time, the clustering framework applies to your entire dataset, creating groups of records that mean similar things.

These results are then grouped in a visualization called a treemap, where clusters represent blocks of different records and are sized according to the number of records in each cluster.

Once your clustering solution has been trained, it can be used as an exploratory tool to identify large groups of records that adhere to common patterns. Clustering is very useful for identifying automation opportunities or reoccurring issues.

Creating a clustering solution

Setting up a clustering solution is somewhat more complex than other models. We'll review the setup and training process here:

1. The first step is to choose a label and name for the clustering solution.
2. Now, you can select a word corpus. This step is similar to what we covered in the *Step 1 – word corpus* section. The word corpus is optional but is likely to produce better results for complex data.
3. Next, you can select the table from which to source the records and the relevant fields on that table to use for the clustering analysis process.

4. The **Group By** selection allows you to apply a logical grouping to your records before the clustering process runs. **Assignment Group**, **Category**, and similar fields that indicate broadly similar families of records are typically used for **Group By** fields.

5. Purity fields can be used to help determine how good your clusters are. Clusters should contain primary records where the purity fields match, so you should select fields that would expect to be the same for similar records.

6. You can select a processing language to process the text for a language in addition to English. This will be useful for implementations that use languages other than English.

7. You can also select stop words, just like in the other solutions. As you can probably tell, maintaining one good stop word list that can be reused between models will be a big time saver relative to building separate stop word lists per model.

8. You can select update and training frequencies that match the rate of change in your data. Consider how often the results will be reviewed and how frequently the analysis is likely to change.

9. Finally, a minimum number of records per cluster can be set. This will depend on your use case, but consider whether a cluster with fewer than 5 to 10 records would be a cluster that provides value in your analysis. You can always reduce the minimum cluster size and retrain it once your largest clusters have been analyzed.

10. Once you've filled in these parameters, you can train your clustering solution. Once it's ready to go, you'll be able to use the **Cluster Visualization** tab to review the results of your clustering solution.

This visualization allows you to interactively filter for cluster sizes and quality. This will allow you to focus on the largest and most consistent clusters first, which are the ones most likely to provide useful ticket groupings. Clicking through the clusters will allow you to see the top 25 representative records, which can be used to identify common root cases and optimization opportunities.

Between similarity, clustering, regression, and classification, ServiceNow provides the tools to address most record-level ML use cases.

Summary

This chapter has given you an overview of AI and ML as it relates to the ServiceNow platform. We've covered features and their applications in terms of creating value, as well as some of the common pitfalls in their application. We explored the differences between the types of predictive intelligence models and the key considerations for their training and use. While the features that were covered in this chapter represent part of the core of AI in ServiceNow and are the most broadly applicable, the ServiceNow platform has two more broadly applicable capabilities that allow us to build incredibly engaging user experiences – NLP and AI search. We'll address them in the next chapter.

11
Designing Exceptional Experiences

So far in this book, we have focused on the parts of ServiceNow that most often form the core of the business case such as how to optimize the processes to maximize business value while avoiding common challenges. One area we have not spent much time on is the creation of an exceptional **user experience** (**UX**) in the platform or the tools that can be used to develop this experience.

In this chapter, we'll cover the parts of the ServiceNow platform that most strongly support an exceptional UX, helping you understand the various options available for building a great UX and how to leverage advanced AI capabilities such as AI Search and **natural language processing** (**NLP**) to provide consumer-grade experiences to your customers. We'll cover such diverse areas as the following:

- Types of ServiceNow user interfaces
- Portals
- Workspaces
- Conversational interfaces

As we did with other chapters, we'll focus on the *what* and the *why* of your implementations, leaving the *how* to ServiceNow's abundant documentation and training programs.

Types of ServiceNow user interfaces

As of 2022, the ServiceNow platform has had three major waves of **user interface** (**UI**) technology deployed and used in the platform. In this section, we'll make you aware of these and their context in the history of ServiceNow to inform your decisions on how you'll develop experiences for your users.

Jelly

We will cover Jelly briefly because, as of the San Diego release and the Polaris UI, the majority of ServiceNow's interface has moved away from this technology. Apache Jelly, or Jelly as it's known in the ServiceNow ecosystem, is a scripting and templating engine that allows for the relatively efficient creation of customized UI elements integrated with the ServiceNow JavaScript engine. For over a decade, Jelly was the main tool in a developer's toolbox for creating UI components in the **Native View** or non-portal view of ServiceNow. Jelly was also the primary tool for creating end user-facing portals until the Geneva release in late 2015. Jelly was a great tool for the time and an important part of ServiceNow's technology stack for over a decade. Today, however, there is virtually no case where a modern customer should need to make use of Jelly for UI configuration or customization except for limited support of legacy use cases that have not yet been modernized. As a rule of thumb, as of late 2022 and for customers on the San Diego release or later, we would recommend extreme caution when any new functionality is suggested to be created in Jelly code. Jelly also had a fully functional end user portal experience called the **Content Management System** (**CMS**) but over time, ServiceNow customers began to call for more flexible and powerful development tools to allow rapid, modern web design to be incorporated into ServiceNow end user experiences.

Service Portal

In response to the need for a more modern and flexible end user experience, ServiceNow developed a way to build AngularJS widgets in ServiceNow and connect them in a modular way to create even more appealing and capable portals for users. ServiceNow also made use of this Angular framework to develop some of the early workspaces in the platform because the technology lends itself well to interactive and dynamic pages.

The vast majority of customer-facing portals in 2022 are still developed using this **Service Portal** technology, which provided fully customizable portals that could be tailored for end users. However, as of the Rome release, an **Employee Center** application is available as the recommended tool for the creation of customer-facing web portals. Employee Center uses the Service Portal technology but provides a far more structured and prescriptive approach than traditional service portals.

Most of the rest of this section will cover the end user experience in the context of the Employee Center journey, as this is currently the recommended end user experience framework and Employee Center provides a general approach to developing an end user portal that does not vary a great deal based on the implementation technology. Before we look at that implementation journey, it's worthwhile to consider the next wave of frontend technology that is likely to become the standard in an upcoming release, which is branded by ServiceNow as the **Next Experience UI**.

Next Experience

Building on the success of ServiceNow's Service Portal, Next Experience is likely to improve the ability of customers to configure web portals for their end users. As of the San Diego release, Next Experience portals can be used as the portal interface for custom applications but are not recommended for building your Enterprise Service portal, although this is expected to change in a subsequent release. As the Next Experience UI is an emerging area, we'll cover the key differences from Service Portal so that you understand when to choose one or the other. Let's look at the application of UI technologies in the context of the major use cases for the ServiceNow UI. We'll start with portal experiences before moving on to the fulfiller experience.

Portals

Portals in ServiceNow are web pages targeted at a broad population within an organization (typically all employees) with the goal of providing effective service and information to a broad customer base. Portals host interactive processes such as requests for goods and services, articles, and informative content as well as outstanding tasks and approvals. Portal content may be the same for all visitors but is often targeted based on information about the user such as country, job function, or other targeting criteria.

Types of portals

End user portals tend to fit into three categories depending on the scope of their coverage and the degree of specialization in their functionality. We can divide portals into three portal types:

- **Enterprise portals**: An enterprise portal serves as a one-stop shop for your organization to access corporate services and can span **Human Resources** (**HR**), **Information Technology** (**IT**), Legal, Finance, and more.

- **Departmental portals**: Departmental portals are focused on providing service and information from a single department or group. An HR portal or an IT portal is a clear example of this portal type.

- **Application portals**: Some ServiceNow custom applications have an end user-facing component that calls for a portal to be created for that specific use case. Examples of application portals can include portals created for third-party applications, external customer support, or tailored interfaces for complex business processes that don't exist in the out-of-the-box processes.

Some portals even evolve through multiple types, often growing from a departmental portal and expanding their scope to other processes. By planning your portal journey correctly from day 1, you can ensure it will scale as your needs expand.

The journey from department to enterprise portals

Let's look at the first two cases together because they currently have a single recommended approach, Employee Center, and because they typically represent a continuous journey from departmental to enterprise portals. It is common for ServiceNow implementations to deploy portals for one or two departments in a major release and then to expand to new use cases and departments in a subsequent release. By adopting Employee Center even where your use case calls for a departmental portal, you will be setting a foundation that will scale to other processes far more efficiently than if you develop a department-specific portal. If your initial release calls for multiple departments to be providing end user services, then the case for the Employee Center approach is only strengthened as it will enable you to deploy each department's processes into a common framework. Once your portal is serving as the enterprise hub for service management, you may even choose to expand that scope to encompass employee engagement and content delivery:

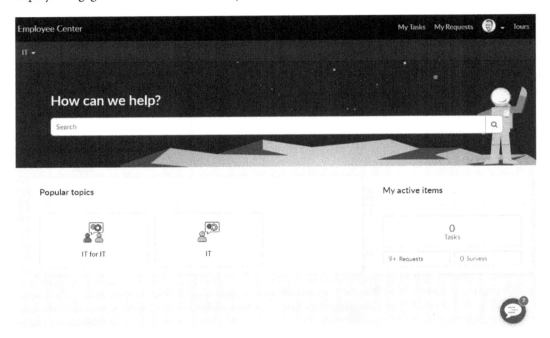

Figure 11.1 – An illustrative Employee Center portal with the default configuration

Employee Center will allow you to realize the vision of a single place to provide services to your employees but the potential doesn't stop there. Let's look at how Employee Center Pro allows you to further increase the impact of your portal.

Employee Center Pro

The final stage in the enterprise portal journey is to go beyond service management and to make Employee Center a destination for your employees to find information, even replacing your corporate intranet. To help facilitate this, ServiceNow offers an Employee Center Pro app as an optional value-added capability that provides enhanced capabilities that allow ServiceNow to act as a corporate content repository in addition to being the service delivery framework. The Pro level of Employee Center adds content publishing, communities, and enterprise-level search capabilities. In your deployment, a good rule of thumb is that if you are aiming to collaboratively engage customers or to implement or replace an Intranet solution, then you should be considering Employee Center Pro.

Content taxonomy

A key concept for deploying Employee Center is the creation of your information taxonomy, which is a structured tree of topics that can be associated with content in your portal. The taxonomy is used to render the menus and topic-based pages and enhance the search experience, so it is worth investing time in creating a logical taxonomy that makes sense for your content. Remember that the creation of content such as catalog items, knowledge articles, and so on should be aligned to value, as we discussed in the opening chapters; in the same way, the taxonomy should serve to provide value to users by helping to organize the items you've created and improving the user's ability to find the right content to solve their specific needs.

> **Important note**
> ServiceNow uses the term *taxonomy* in multiple contexts. There is the Service Portfolio Management taxonomy, GRC taxonomy, and Employee Center taxonomy. The content in this section refers only to the Employee Center taxonomy. It is particularly important when consulting the documentation site and system menus to ensure you're looking at the right taxonomy.

ServiceNow provides an Employee Experience Taxonomy out of the box that will be populated with topics based on your installed applications. It is suggested to clone this taxonomy as a starting point for your own taxonomies and modify it as needed. Once your taxonomy has been defined, you should consider the visual presentation of the content. In the next section, we will look at tailoring the look and feel of your portal.

Custom app portals and portal customizations

ServiceNow's portal experiences can be fine-tuned to meet the visual and functional needs of your organization. This can range from simple color updates to modifying the structure and behavior of pages and their component widgets.

Many customers choose to modify and tailor their Employee Center portal's style using Service Portal's Branding Editor to adjust the look and feel by updating the colors of the portal theme to reflect their own color palette. This is useful if you want to give your portal a look that matches your corporate identity with brand-approved colors and your own organizational or portal logo:

Figure 11.2 – Service Portal's Branding Editor

Using this branding editor, you will be able to navigate through the various pages in your portal to see how the changes appear on each page. This will help identify any changes that need to be made or to confirm that the design is ready to be used.

Employee Center also allows for the configuration of the items that are displayed in the out-of-the-box widgets, including the following items:

- To-dos
- Available requests
- Chat functionality
- Recommendations
- Active items
- Cross-channel favorites
- Launchable business applications
- Employee profiles
- Custom footers

These configurations allow you to fine-tune the behavior of the portal and act in conjunction with your content taxonomy to form the basis of your portal experience.

If you wish to make further functional changes to the Employee portal, you can use the Service Portal configuration tools such as the Designer, Page Editor, and Widget Editor (as well as the backend forms for these objects) but this should only be done with an abundance of caution and a strong business case. Employee Center continues to be invested in by ServiceNow and modifications to out-of-the-box capabilities can lead to you missing out on new features or issues during upgrades.

Application portals

These warnings against customization are appropriate when considering a modification to an out-of-the-box portal such as Employee Center but, in some cases, you may choose to develop a new portal for a custom application. This is where the third category of **application portals** comes in. These portals are intended to facilitate more complex processes where you have established a clear business need. As of San Diego, the recommended approach for custom application portals is to use the portal **Experience** option in App Engine Studio, which uses the Next Experience UI framework as its UI technology:

Figure 11.3 – Adding an experience from the App Engine Studio Experience tab

When adding an experience in App Engine Studio, you can select from some pre-existing experience templates. For most end user use cases you'll find the **Portal** experience to give you the best starting point.

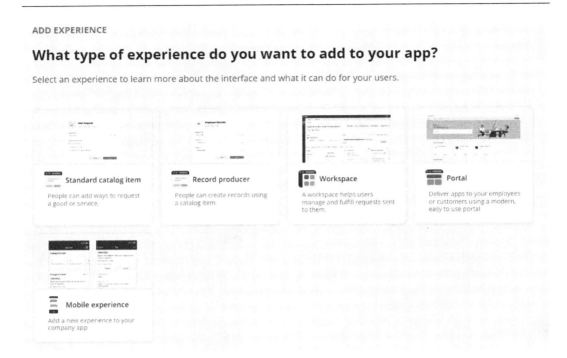

Figure 11.4 – A Portal experience can be selected in App Engine
Studio when building a custom application

It's also still possible to create Service Portal-based custom application portals, which is a choice that some customers might make if the availability of out-of-the-box Service Portal widgets suitable to their use cases would drastically simplify portal development when compared to the efforts using UI Builder.

Portals provide an excellent way for your organization to interact with ServiceNow-enabled processes, but they are used by the *consumers* of services rather than the *providers* of services. To provide maximum value, ServiceNow also provides a strong backend UI capability that helps your teams to deliver these services more efficiently and effectively.

In the next section, we'll explore the interfaces that can be used to get work done most efficiently within ServiceNow and to help you navigate the decisions you'll need to make when deploying workspaces for your fulfillment teams.

Workspaces

The most basic interface in ServiceNow is the native view, which consists of the forms and lists that make up the core of ServiceNow's traditional UI. While it is possible, in most cases, to use this ServiceNow native view to complete almost all tasks that are required to provide service and maintain operations, the use of the native view is less efficient than a specially designed type of interface known as a workspace.

Workspaces have been developed specifically to optimize service delivery and represent the most efficient way for Level 1 and 2 agents to engage with the platform. You should note that administrators and power users will continue to use the native view, as many of the configuration and administration features are only exposed in the native view.

Until the Next Experience UI framework is fully integrated into all ServiceNow product workspaces, there are two types of workspaces – **Legacy Workspaces** (also called agent workspaces) and **Configurable Workspaces** (sometimes referred to as WEP workspaces in ServiceNow documentation).

New workspaces in the Now Platform will be Configurable Workspaces and use the Next Experience framework and can be tailored in UI Builder. You are also able to develop your own workspaces using UI Builder. Since there are still some workspace experiences that have not been moved over to the Next Experience framework, in some cases, you will have the choice of a Legacy Workspace or a Configurable Workspace, so it is useful to understand the difference and when to use each.

Determining which workspaces to use

Legacy Workspaces are most notably still used for ITSM and ITOM service delivery as of the San Diego release of ServiceNow, while CSM, vendor management, cloud migration, and several other workspaces are already available as Configurable Workspaces, with more expected as store applications and in future releases.

For the time being, Legacy Workspaces may continue to be used by organizations that have not yet migrated but for new implementations, it would be best to use a Configurable Workspace if it were available. Developing and customizing Legacy Workspaces is not well supported, so all new workspaces developed as part of a custom application should be created using Next Experience and UI Builder.

Because all ServiceNow user interfaces operate on a unified data model, it is possible and relatively seamless to use different workspace types together, even where processes interact. For example, a CSM Configurable Workspace can be used to serve customer cases, while incidents created from cases can continue to be worked via the Legacy ITSM Agent Workspace. Configurable Workspaces will tend to offer far greater versatility to be configured to meet specific business needs using UI Builder.

Just as it is recommended to think carefully before making changes to a Service Portal, you should carefully consider the value and cost of changes to Configurable Workspaces; when the value justifies the effort to build and maintain the changes, you will be able to make the changes using UI Builder.

One of the advantages of the workspaces is their ability to integrate AI recommendations and search into the context of service delivery. In the previous chapter, we addressed the Predictive Intelligence framework, which powers intelligent recommendations in the workspaces. Specifically, when you are viewing a ticket record in a workspace, Predictive Intelligence can find similar records and even provide field predictions to reduce manual effort and errors. Another ServiceNow AI capability that plays a part in the workspaces is the AI Search capability, which you can tailor to surface relevant content in the context of your workspaces even without UI customization. AI Search is just one of the advanced AI capabilities that can be used to help shape a better user experience for both agents and fulfiller users; in the next section, we'll look at some other examples.

Optimizing experiences with user-facing AI

We have covered the various types of user experiences and outlined the approaches you should take to deploying them in light of the changing landscape of ServiceNow UI technology so far. In this section, we will cover some of the supporting but highly value-generating features of the ServiceNow platform that allow us to shape experience, not only through design but also through intelligence.

AI Search

The AI Search capability in ServiceNow is another multipurpose solution that is useful in improving the search experience of users in the Employee Center portal as well as surfacing relevant information to agents working in their respective workspaces.

AI Search utilizes sophisticated algorithms to find and rank information in a way that generates more meaningful results than prior generations of search technology. Specifically, AI Search utilizes a form of language processing known as **semantic search** and enhances this with a ranking model that combines administrator input and aggregated user behavior to determine the correct results to display and the order in which to display them.

AI Search configurability

When configuring AI Search, you can specify the content that should be indexed and various other parameters, such as stop words and boosting and blocking rules that help surface the most relevant content. You can also configure a special type of result known as a **Genius Result**, which can be triggered with the ServiceNow **Natural Language Understanding** (NLU) features.

Genius Results should be generated only in the case that the results are most strongly relevant to the user's query. Some good examples of this would be when a search can return a closely related catalog item or highly relevant knowledge article. In the language of the previous chapter, Genius Results have very low coverage but high precision, meaning that a Genius Result is unlikely to be provided for any given search term, but when shown, a Genius Result is very likely to be correct. You can create Genius Results from records in ServiceNow or from other available data sources accessible via an API.

AI Search works with the help of an external search service that runs alongside your ServiceNow instance and provides a consumer-grade search capability in the portal, workspaces, and via APIs that allow access to custom applications and features. AI Search can index content both inside and outside of the Now Platform and offers seamless integration to both portal and workspace experiences. In addition to the presentation through these interfaces, you can use AI Search in a Virtual Agent conversation, which brings us to another form of user experience that is often embedded in ServiceNow's portals, Virtual Agent.

Conversational interfaces

One capability that can provide a lot of value is the Virtual Agent chatbot capability that ServiceNow offers alongside its NLP engine. Virtual Agent allows users to interact via text chat with an automated system to quickly handle simple tasks, even when no human agents are available. If a Virtual Agent is unable to handle a query, then it can be transferred to a human who can carry on the support conversation using the chat interface built into the agent workspaces.

Incorporating Virtual Agent into your user experience

Virtual Agent is typically presented through a service portal in the form of a webchat window. Virtual Agent serves as a highly automated first line of triage and can address user needs that would otherwise have required the services of a human agent.

Virtual Agents can also be exposed through an external chat channel such as Microsoft Teams, Slack, the ServiceNow mobile application, and even custom conversational interfaces.

When introducing Virtual Agents into your environment, you should be careful to consider how a chat-based channel provides different user experience trade-offs relative to a form-based interface. There are both advantages and disadvantages, and not all use cases lend themselves well to Virtual Agents.

Advantages of Virtual Agent as a user interface

Just like a conversation with a regular human, conversational interfaces such as ServiceNow Virtual Agent can take different paths through a conversation in response to the choices of the user. This allows for a highly dynamic experience that reacts to the user and asks only for information that's not already available in the system or in the user's past inputs. Virtual Agent topics can be configured to take action automatically based on the user inputs, often orchestrating work inside of ServiceNow or using external integrations; this allows a user to get an instantaneous resolution to their needs and results in both improved customer satisfaction and lower support costs because a human agent was not required.

Disadvantages of conversational interfaces

Virtual Agent interfaces do have some significant drawbacks relative to a standard web portal. The logic behind Virtual Agent conversations is specified by the topic developers and follows predetermined paths, unlike a human conversation. If you're not careful about the design of these paths, the user can feel locked into an incorrect path and get frustrated with the experience.

The chat interface is also low-bandwidth, meaning that you're limited in the amount of information you can convey over the chat channel. The chatbot interfaces, such as Virtual Agent, require you to think carefully about what information you want to present to the user with each new message.

Recommendations for Virtual Agent

The following recommendations will help you make the most of your Virtual Agent deployment and to create more value than just a portal alone:

- Consider enabling topic switching when the user is asked for input, as this enables a user to get away from a "wrong" path as early as possible to reduce frustration. If agents are available, the option to chat with a human operator is an excellent fallback. Virtual Agents should be seen as the first line of support (often called Tier 0), not your only support offering.

- When asking for user input, filter the options to as small a list as possible to minimize the effort for the user. As a rule, consider five to seven items as a maximum number of options per interaction. If you need more, you may prefer a text input with NLP enabled.

- Reusable topic blocks can be used to create consistency and reusability across multiple conversations. Remember the goal of the Virtual Agent is to get things done quickly and without human interaction and that a familiar experience contributes to that goal. The out-of-the-box topic blocks can also be composed into your topics to provide a rich capability very quickly.

- Not every form is appropriate for a Virtual Agent conversation. When a large amount of user input is required, the chat interface can feel clunky. Users are comfortable tabbing or clicking through forms and may want to review all their responses before submitting.

- Where information or inputs are very complex, a Virtual Agent can also serve to find the appropriate resource and link to it for the user rather than forcing that information to flow through the chat channel. If your friends or family wanted to share a long news article with you, they would send you a link, not copy the article text into their messages to you. Consider taking the same approach here.

- Surfacing the Virtual Agent through a common organizational channel such as Teams or Slack can save the user the effort of navigating to the portal and cut down friction in the process. Also, consider that these channels make the Virtual Agent available on mobile devices, expanding their reach significantly.

Virtual Agents are made much more powerful and useful when integrated with the ability to understand written language. Let's look at ServiceNow's language processing features.

ServiceNow NLU

When a user first engages a Virtual Agent in a conversation, the task of the chatbot is to match the user's needs to a defined topic or a fallback. This can be accomplished with basic keyword matching or by using ServiceNow's own implementation of NLU. NLU models can be trained directly in ServiceNow and broadly serve two main purposes – **intent recognition** and **entity recognition**.

Intent recognition is the task of taking some input sentence and mapping it to a predefined intention that can trigger a specific flow. An input sentence might read *"Hey there ServiceNow, I'd like to order a new laptop."* This could be mapped to an *Order Hardware* intent. Note that the actual intents do vary from customer to customer so one organization may match this text to the *Order Something* intent while another might map it to *Request Replacement Laptop*. A good language model is one that most reliably finds the intents closest to the user's requests and one that does not often trigger the wrong topic flow.

Entity recognition is a second function of the ServiceNow language models and allows us to recognize not just the type of topic but even some specific data that might be useful in acting on that intention. If a user was to say to the Virtual Agent *"Please order me a black iPhone 13 Pro,"* the system could not only detect a request for a new phone but also key facts that we might otherwise have had to prompt the user for, such as the model (*iPhone 13 Pro*) and the phone color (*black*).

By combining NLU and Virtual Agent, you can allow users to provide a great deal of information in a short message that puts you well on your way to intelligently responding to their query in an automated way.

Summary

In this chapter, we've covered the major types of user interfaces in ServiceNow and addressed portals, workspaces, and the supporting AI technologies that make these experiences more relevant and intelligent. Throughout the book, we've aimed to create a useful store of information that applies across releases but in this chapter, we've had to get into the details of an important shift in the ServiceNow UI that makes some of the advice necessarily specific to the version available as of the time of writing this book., By explaining the motivations and technologies involved, you will now have the tools to make similar decisions well into the future as well.

With this chapter, we also conclude the final section of this book in which we've looked at how to drive innovation on Now Platform and to create efficient, intelligent experiences for end users and agents alike.

With the foundation you've gained in this book, we hope you'll be better prepared for your journey to make an impact with ServiceNow. From this point on, you should prepare for those capabilities most likely to deliver value and be useful to you in your roles.

No matter where you are within the ServiceNow ecosystem, you will benefit from getting hands-on with the software. Go to `https://developer.servicenow.com` and sign up for a personal Developer Instance where you can build and explore on your own.

For those pursuing a more technical path as a developer, the book *ServiceNow Development Handbook* (Tim Woodruff) provides excellent implementation tips.

For those focused on product and process, the ServiceNow documentation for each product area remains the most complete and exhaustive guide to the product and best practices.

Finally, ServiceNow has a vibrant community at `https://community.servicenow.com`, where a global community of your ServiceNow peers is willing to help answer questions and offer advice.

It's almost impossible to learn about everything there is to know about ServiceNow given how broad the capabilities are but if you start somewhere, keep learning, and always stay focused on delivering value, you'll most likely be on the right track.

Index

`Packt.com`

Subscribe to our online digital library for full access to over 7,000 books and videos, as well as industry leading tools to help you plan your personal development and advance your career. For more information, please visit our website.

Why subscribe?

- Spend less time learning and more time coding with practical eBooks and Videos from over 4,000 industry professionals

- Improve your learning with Skill Plans built especially for you

- Get a free eBook or video every month

- Fully searchable for easy access to vital information

- Copy and paste, print, and bookmark content

Did you know that Packt offers eBook versions of every book published, with PDF and ePub files available? You can upgrade to the eBook version at `packt.com` and as a print book customer, you are entitled to a discount on the eBook copy. Get in touch with us at `customercare@packtpub.com` for more details.

At `www.packt.com`, you can also read a collection of free technical articles, sign up for a range of free newsletters, and receive exclusive discounts and offers on Packt books and eBooks.

Other Books You May Enjoy

If you enjoyed this book, you may be interested in these other books by Packt:

Driving DevOps with Value Stream Management

Cecil 'Gary' Rupp

ISBN: 9781801078061

- Integrate Agile, systems thinking, and lean development to deliver customer-centric value
- Find out how to choose the most appropriate value stream for your initial and follow-on VSM projects
- Establish better flows with integrated, automated, and orchestrated DevOps and CI/CD pipelines
- Apply a proven eight-step VSM methodology to drive lean IT value stream improvements
- Discover the key strengths of modern VSM tools and their customer use case scenarios
- Understand how VSM drives DevOps pipeline improvements and value delivery transformations across enterprises

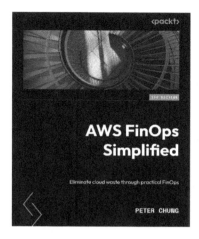

AWS FinOps Simplified

Peter Chung

ISBN: 9781803247236

- Use AWS services to monitor and govern your cost, usage, and spend
- Implement automation to streamline cost optimization operations
- Design the best architecture that fits your workload and optimizes on data transfer
- Optimize costs by maximizing efficiency with elasticity strategies
- Implement cost optimization levers to save on compute and storage costs
- Bring value to your organization by identifying strategies to create and govern cost metrics

Packt is searching for authors like you

If you're interested in becoming an author for Packt, please visit authors.packtpub.com and apply today. We have worked with thousands of developers and tech professionals, just like you, to help them share their insight with the global tech community. You can make a general application, apply for a specific hot topic that we are recruiting an author for, or submit your own idea.

Share Your Thoughts

Now you've finished *ServiceNow for Architects and Project Leaders*, we'd love to hear your thoughts! Scan the QR code below to go straight to the Amazon review page for this book and share your feedback or leave a review on the site that you purchased it from.

https://packt.link/r/1803245298

Your review is important to us and the tech community and will help us make sure we're delivering excellent quality content.

Download a free PDF copy of this book

Thanks for purchasing this book!

Do you like to read on the go but are unable to carry your print books everywhere?

Is your eBook purchase not compatible with the device of your choice?

Don't worry, now with every Packt book you get a DRM-free PDF version of that book at no cost.

Read anywhere, any place, on any device. Search, copy, and paste code from your favorite technical books directly into your application.

The perks don't stop there, you can get exclusive access to discounts, newsletters, and great free content in your inbox daily

Follow these simple steps to get the benefits:

1. Scan the QR code or visit the link below

https://packt.link/free-ebook/978-1-80324-529-4

2. Submit your proof of purchase
3. That's it! We'll send your free PDF and other benefits to your email directly

www.ingramcontent.com/pod-product-compliance
Lightning Source LLC
Chambersburg PA
CBHW060540060326
40690CB00017B/3560